Student Solutions Manual
to accompany

Analytical Chemistry

An Introduction

seventh edition

DOUGLAS A. SKOOG
Stanford University

DONALD M. WEST
San Jose State University

F. JAMES HOLLER
University of Kentucky

STANLEY R. CROUCH
Michigan State University

Australia • Brazil • Japan • Korea • Mexico • Singapore • Spain • United Kingdom • United States

Student Solutions Manual to accompany: Analytical Chemistry: An Introduction, Seventh Edition

Douglas A. Skoog, Donald M. West, F. James Holler, Stanley R. Crouch

ISBN-13: 978-0-03-023492-7

ISBN-10: 0-03-023492-1

Brooks/Cole
10 Davis Drive
Belmont CA 94002-3098
USA

Cengage Learning is a leading provider of customized learning solutions with office locations around the globe, including Singapore, the United Kingdom, Australia, Mexico, Brazil, and Japan. Locate your local office at **international.cengage.com/region**

Cengage Learning products are represented in Canada by Nelson Education, Ltd.

For your course and learning solutions, visit **academic.cengage.com**

Purchase any of our products at your local college store or at our preferred online store **www.ichapters.com**

Printed in the United States of America
5 6 7 11 10 09 08

PREFACE

This Student Solutions Manual contains the answers and solutions to problems marked with an asterisk in **Analytical Chemistry: An Introduction, 7th edition**. For some problems that require essentially identical solutions, we have written a single solution and provided an appropriate cross-reference along with the answer. In regions of titration curves where the calculations are similar, we have written a single solution and provided tables of results. For problems requiring spreadsheets, we have provided copies of Microsoft®Excel spreadsheet solutions.

We have made every effort to eliminate errors in these solutions, but in the event that you discover any discrepancies, please contact us at holler@pop.uky.edu.

December 1999

Douglas A. Skoog
Donald M. West
F. James Holler
Stanley R. Crouch

CONTENTS

Chapter 3

3-1. **(a)** The *millimole* is the amount of an elementary species, such as an atom, an ion, a molecule, or an electron. A millimole contains

$$6.02\times10^{23}\,\frac{\text{particles}}{\cancel{\text{mole}}}\times10^{-3}\,\frac{\cancel{\text{mole}}}{\text{millimole}} = 6.02\times10^{20}\,\frac{\text{particles}}{\text{millimole}}$$

(c) The millimolar mass of a species is the mass in grams of one millimole of the species.

3-3. $1\ \text{L} = 10^{-3}\ \text{m}^3$

$$1\ \text{M} = \frac{1\ \text{mol}}{\text{L}} = \frac{1\ \text{mol}}{10^{-3}\ \text{m}^3}$$

$1\ \text{Å} = 10^{-10}\ \text{m}$

3-4. **(a)**

$$2.5\times10^4\ \text{Hz}\times10^3\,\frac{\text{k Hz}}{\text{Hz}} = \underline{\underline{25\ \text{k Hz}}}$$

(c)

$$8.43\times10^5\ \mu\text{mol}\times10^{-3}\,\frac{\text{mmol}}{\mu\text{mol}} = \underline{\underline{843\ \text{mmol}}}$$

(e)

$$7.44\times10^4\ \text{nm}\times10^{-3}\,\frac{\mu\text{m}}{\text{nm}} = \underline{\underline{74.4\ \mu\text{m}}}$$

3-5.

$$4.62\ \text{g Na}_3\text{PO}_4\times\frac{1\ \text{mol Na}_3\text{PO}_4}{163.94\ \text{g}}\times\frac{3\ \text{mol Na}^+}{1\ \text{mol Na}_3\text{PO}_4}\times\frac{6.02\times10^{23}\ \text{Na}^+\ \text{ion}}{\text{mol Na}^+} = \underline{\underline{5.09\times10^{22}\ \text{Na}^{2+}\,\text{ion}}}$$

3-7. **(a)**

$$6.84\ \text{g B}_2\text{O}_3\times\frac{1\ \text{mol B}_2\text{O}_3}{69.62\ \text{g B}_2\text{O}_3} = \underline{\underline{0.0982\ \text{mol B}_2\text{O}_3}}$$

(b)

$$296\ \text{mg}\times\frac{1\ \text{g}}{1000\ \text{mg}}\times\frac{1\ \text{mol}}{381.37\ \text{g}} = \underline{\underline{7.76\times10^{-4}\ \text{mol Na}_2\text{Ba}_4\text{O}_7\cdot10\text{H}_2\text{O}}}$$

(c)

$$8.75\ \text{g Mn}_3\text{O}_4\times\frac{1\ \text{mol Mn}_3\text{O}_4}{228.81\ \text{g Mn}_3\text{O}_4} = \underline{\underline{0.0382\ \text{mol Mn}_3\text{O}_4}}$$

(d) $$67.4 \text{ mg CaC}_2\text{O}_4 \times \frac{1 \text{ g}}{1000 \text{ mg}} \times \frac{1 \text{ mol CaC}_2\text{O}_4}{128.10 \text{ g CaC}_2\text{O}_4} = \underline{\underline{5.26 \times 10^{-4} \text{ mol CaC}_2\text{O}_4}}$$

3-9. (a) $$2.00 \text{ L} \times 10^3 \frac{\text{mL}}{\text{L}} \times \frac{2.76 \times 10^{-3} \text{ mmol KMnO}_4}{\text{mL}} = \underline{\underline{5.52 \text{ mmol KMnO}_4}}$$

(b) $$750 \text{ mL} \times 0.0416 \frac{\text{mmol KSCN}}{\text{mL}} = \underline{\underline{31.2 \text{ mmol KSCN}}}$$

(c) $$\frac{4.20 \text{ mg CuSO}_4}{\text{L}} \times \frac{1 \text{ mmolCuSO}_4}{159.61 \times 10^3 \text{ mg CuSO}_4} \times 0.250 \text{ L} = \underline{\underline{6.58 \times 10^{-3} \text{ mmol}}}$$

(d) $$3.50 \text{ L} \times 0.276 \frac{\text{mol}}{\text{L}} \times 10^3 \frac{\text{mmol}}{\text{mol}} = \underline{\underline{966 \text{ mmol}}}$$

3-11. (a) $$0.666 \text{ mol HNO}_3 \times 63.013 \frac{\text{g HNO}_3}{\text{mol}} \times 1000 \frac{\text{mg}}{\text{g}} = \underline{\underline{4.20 \times 10^4 \text{ mg HNO}_3}}$$

(b) $$300 \text{ mL MgO} \times 40.30 \frac{\text{mg MgO}}{\text{mmol}} = \underline{\underline{1.21 \times 10^4 \text{ mg MgO}}}$$

(c) $$19.0 \text{ mol} \times 80.043 \frac{\text{g NH}_4\text{NO}_3}{\text{mol}} \times \frac{10^3 \text{ mg}}{\text{g}} = \underline{\underline{1.52 \times 10^6 \text{ mg NH}_4\text{NO}_3}}$$

(d) $$5.32 \text{ mol} \times 548.23 \frac{\text{g}}{\text{mol}} \times \frac{10^3 \text{ mg}}{\text{g}} = \underline{\underline{2.92 \times 10^6 \text{ mg}}}$$

3-13. (a) $$26.0 \text{ mL} \times 0.150 \frac{\text{mmol}}{\text{mL}} \times 342 \frac{\text{mg}}{\text{mmol}} = \underline{\underline{1.33 \times 10^3 \text{ mg sucrose}}}$$

(b) $$2.92 \text{ L} \times 5.23 \times 10^{-3} \frac{\text{mol H}_2\text{O}_2}{\text{L}} \times 34.015 \frac{\text{g}}{\text{mol}} \times 10^3 \frac{\text{mg}}{\text{g}} = \underline{\underline{519 \text{ mg H}_2\text{O}_2}}$$

3-14. (a) $$0.450 \text{ L} \times 0.164 \frac{\text{mol}}{\text{L}} \times 34.015 \frac{\text{g H}_2\text{O}_2}{\text{mol}} = \underline{\underline{2.51 \text{ g H}_2\text{O}_2}}$$

(b) $27.0\text{ mL} \times 8.75 \times 10^{-4} \dfrac{\text{mmol}}{\text{mL}} \times 0.122 \dfrac{\text{g HBz}}{\text{mmol}} = \underline{\underline{2.88 \times 10^{-3}\text{ g HBz}}}$

3-15. **(a)** $[Na^+] = 0.0235 + 0.0503 = 0.0738$ $\quad$ $pNa = -\log 0.0738 = \underline{\underline{1.132}}$

$pCl = -\log 0.0235 = \underline{\underline{1.629}}$

$pOH = -\log 0.0503 = \underline{\underline{1.298}}$

(c) $pH = -\log 0.800 = \underline{\underline{0.097}}$

$pCl = -\log (0.800 + 2 \times 0.101) = \underline{\underline{-0.001}}$

$pZn = -\log 0.101 = \underline{\underline{0.99}}$

(e) $pK = -\log (4 \times 3.79 \times 10^{-7} + 4.12 \times 10^{-7}) = \underline{\underline{5.715}}$

$pOH = -\log 4.12 \times 10^{-7} = \underline{\underline{6.385}}$

$pFe(CN)_6 = -\log 3.79 \times 10^{-7} = \underline{\underline{6.421}}$

3-16. **(a)** $[H_3O^+] = 10^{-9.21} = \underline{\underline{6.2 \times 10^{-10}\text{ M}}}$

(c) $[H_3O^+] = 10^{-0.45} = \underline{\underline{3.5 \times 10^{-1}\text{ M}}}$

(e) $[H_3O^+] = 10^{-7.32} = \underline{\underline{4.8 \times 10^{-8}\text{ M}}}$

(g) $[H_3O^+] = 10^{+0.21} = \underline{\underline{1.6\text{ M}}}$

3-17. **(a)** $pNa = pBr = -\log 0.0100 = \underline{\underline{2.000}}$

$pH = pOH = 7.00$

(c) $pBa = -\log 3.5 \times 10^{-3} = \underline{\underline{2.46}}$

$$\text{pOH} = -\log 7.0 \times 10^{-3} = \underline{\underline{2.15}}$$

$$\text{pH} = 14.00 - 2.15 = 11.85$$

(e) $$\text{pCa} = -\log 5.2 \times 10^{-3} = \underline{\underline{2.28}}$$

$$\text{pBa} = -\log 3.6 \times 10^{-3} = \underline{\underline{2.44}}$$

$$\text{pCl} = -\log (10.4 \times 10^{-3} + 7.2 \times 10^{-3} = \underline{\underline{1.75}}$$

$$\text{pH} = \text{pOH} = 7.00$$

3-18. (a) $$[H_3O^+] = 10^{-8.67} = \underline{\underline{2.1 \times 10^{-9}\ \text{M}}}$$

(c) $$[Br^-] = 10^{-0.034} = \underline{\underline{0.925\ \text{M}}}$$

(e) $$[Li^+] = 10^{+0.321} = \underline{\underline{2.09\ \text{M}}}$$

(g) $$[Mn^{2+}] = 10^{-0.0025} = \underline{\underline{0.99\ \text{M}}}$$

3-19. (a) $$\frac{1080\ \text{mg Na}^+}{\text{kg water}} \times \frac{1.02\ \text{kg}}{\text{L}} \times \frac{\text{mol Na}^+}{2.299 \times 10^4\ \text{mg}} = \underline{\underline{4.79 \times 10^{-2}\ \text{M Na}^+}}$$

$$\frac{270\ \text{mg SO}_4^{2-}}{\text{kg water}} \times \frac{1.02\ \text{kg}}{\text{L}} \times \frac{\text{mol SO}_4^{2-}}{9.606 \times 10^4\ \text{mg}} = \underline{\underline{2.87 \times 10^{-2}\ \text{M SO}_4^{2-}}}$$

(b) $$\text{pNa} = -\log 4.79 \times 10^{-2} = \underline{\underline{1.320}}$$

$$\text{pSO}_4 = -\log 2.87 \times 10^{-3} = \underline{\underline{2.543}}$$

3-21. (a) $$c = \frac{6.34\ \text{g} \times \frac{1\ \text{mol}}{277.85\ \text{g}}}{2.000\ \text{L}} = \underline{\underline{1.141 \times 10^{-2}\ \text{M salt}}}$$

(b) $$[Mg^{2+}] = c = \underline{\underline{1.141 \times 10^{-2}\ \text{M}}}$$

(c) $[Cl^-] = 3c = \underline{\underline{3.423 \times 10^{-2}\ M}}$

(d) $$\frac{6.34\ g}{2.000\ L} \times \frac{1\ L}{1000\ mL} \times 100\% = \underline{\underline{0.317\%\ (w/v)}}$$

(e) $$25.0\ mL \times 3.423 \times 10^{-2}\,\frac{mmol}{mL} = \underline{\underline{0.856\ mmol\ Cl^-}}$$

(f) $$1.141 \times 10^{-2}\,\frac{mol\ salt}{L} \times \frac{1\ mol\ K^+}{mol\ salt} \times \frac{39.1\ mg\ K^+}{mmol} \times 1000\,\frac{mL}{L} = \underline{\underline{446\ ppm}}$$

(g) $pMg = -\log 1.141 \times 10^{-2} = \underline{\underline{1.943}}$

(h) $pCl = -\log 3.423 \times 10^{-2} = \underline{\underline{1.466}}$

3-23. **(a)** $$1.059\,\frac{g}{mL} \times \frac{6.42\ g\ Fe(NO_3)_3}{100\ g} \times \frac{1\ mol\ Fe(NO)_3}{241.8\ g} \times \frac{10^3\ mL}{L} = \underline{\underline{0.281\ M\ Fe(NO)_3}}$$

(b) $$0.281\ M\ Fe(NO)_3 \times \frac{3\ mol\ NO_2^-}{mol\ Fe(NO_3)_3} = \underline{\underline{0.843\ M\ NO_3^-}}$$

(c) $$1.059\,\frac{g}{mL} \times 10^3\,\frac{mL}{L} \times 6.42\ g\,\frac{Fe(NO_3)_3}{100\ mL} = \underline{\underline{68.0\,\frac{g\ Fe(NO_3)_3}{L\ soln}}}$$

3-25. $$500\ mL\ EtOH \times \frac{4.75\ g\ EtOH}{100\ mL} = 23.8\ g\ EtOH$$

(a) Dissolve 23.8 g of EtOH in water and dilute to 500 mL.

(b) Mix 23.8 g EtOH with 476.2 g H_2O.

(c) Dissolve 23.8 mL of EtOH and dilute to 500 mL.

3-27. $$c_{H_3PO_4} = \frac{1.69 \times 10^3\ g\ reagent}{L\ reagent} \times \frac{85\ g\ H_3PO_4}{100\ g\ reagent} \times \frac{1\ mol\ H_3PO_4}{98.0\ g\ H_3\ PO_4} = \frac{14.66\ mol\ H_3PO_4}{L\ reagent}$$

$$\text{Amount } H_3PO_4 \text{ needed} = 0.750\ L \times 6.00\,\frac{mol}{L} = 4.50\ mol\ H_3PO_4$$

$$\text{Vol concd reagent} = \frac{4.50 \text{ mol } H_3PO_4}{14.66 \text{ mol } H_3PO_4/\text{L}} = 0.307 \text{ L}$$

Dilute 307 mL reagent to 750 mL.

3-29. (a)

$$500 \text{ mL} \times 0.0650 \frac{\text{mmol AgNO}_3}{\text{mL}} \times 0.16987 \frac{\text{g AgNO}_3}{\text{mmol}} = 5.52 \text{ g}$$

Dissolve 5.52 g $AgNO_3$ in water and dilute to 500 mL.

(b)

$$1000 \text{ mL} \times 0.285 \frac{\text{mol HCl}}{\text{mL}} = \text{mL concd reagent} \times 6.00 \frac{\text{mmol}}{\text{mL}}$$

mL concd reagent = 47.5

Dilute 47.5 mL of 6.00 M HCl to 1.00 L.

(c)

$$0.400 \text{ L} \times 0.0875 \frac{\text{mol K}^+}{\text{L}} \times \frac{1 \text{ mol K}_4\text{Fe(CN)}_6}{4 \text{ mol K}^+} \times 368.35 \frac{\text{g K}_4\text{Fe(CN)}_6}{\text{mol K}_4\text{Fe(CN)}_6} = 3.22 \text{ g}$$

Dissolve 3.22 g $K_4Fe(CN)_6$ in water and dilute to 400 mL.

(d)

$$600 \text{ mL} \times \frac{3.00 \text{ gBaCl}_2}{100 \text{ mL}} \times \frac{1 \text{ mol BaCl}_2}{208.23 \text{ g}} \times \frac{1 \text{ L}}{0.400 \text{ mol}} = 0.216 \text{ L} \quad \text{or} \quad 216 \text{ mL}$$

Dilute 216 mL of $BaCl_2$ solution to 600 mL.

(e)

$$c_{\text{HClO}_4} = 1.60 \frac{\text{g}}{\text{mL}} \times \frac{60 \text{ g HClO}_4}{100 \text{ g}} \times \frac{1 \text{ mmol}}{0.10046 \text{ g}} = 9.56 \frac{\text{mmol HClO}_4}{\text{mL}}$$

$$2000 \text{ mL HClO}_4 \times 0.120 \frac{\text{mol}}{\text{mL}} = \text{mL concd reagent} \times 9.56 \frac{\text{mmol HClO}_4}{\text{mL}}$$

$$\text{mL concd reagent} = \frac{2000 \times 0.120}{9.56} = 2.51 \text{ mL}$$

Dilute 25.1 mL of the concentrated reagent to 2.00 L.

(f)
$$9.00 \times 10^3 \text{ mL soln} \times 1.00 \frac{\text{g soln}}{\text{mL soln}} \times \frac{60 \text{ g Na}^+}{10^6 \text{ g soln}} \times \frac{1 \text{ mol Na}^+}{22.99 \text{ g Na}^+} \times$$

$$\frac{1 \text{ mol Na}_2\text{SO}_4}{2 \text{ mol Na}^+} \times \frac{142.0 \text{ g Na}_2\text{SO}_4}{\text{mol Na}_2\text{SO}_4} = 1.67 \text{ g Na}_2\text{SO}_4$$

Dissolve 1.67 g Na_2SO_4 in water and dilute to 9.00 L.

3-31.
$$50.0 \text{ mL} \times 0.250 \frac{\text{mmol La}^{3+}}{\text{mL}} = 12.5 \text{ mmol La}^{2+}$$

$$75.0 \text{ mL} \times 0.302 \frac{\text{mmol IO}_3^-}{\text{mL}} = 22.65 \text{ mmol IO}_3^-$$

Thus, La^{3+} is in excess and the amount $La(IO)_3^-$ produced = 22.65/3 = 7.55 mmol.

$$7.55 \text{ mmol La(IO}_3^-)_3 \times 0.6636 \frac{\text{g La(IO}_3)_3}{\text{mmol}} = \underline{\underline{5.01 \text{ g La(IO}_3)_3}}$$

3-33. $Na_2CO_3 + 2HCl \rightarrow 2NaCl + H_2O + CO_2$

$$0.1120 \text{ g} \times \frac{1 \text{ mol}}{105.99 \text{ g}} = 1.057 \times 10^{-3} \text{ mol Na}_2\text{CO}_3 = 1.057 \text{ mmol Na}_2\text{CO}_3$$

$$100.0 \text{ mL} \times 0.0497 \text{ M} = 4.97 \text{ mmol HCl}$$

$$4.97 \text{ mmol HCl} \times \frac{1 \text{ mmol Na}_2\text{CO}_3}{2 \text{ mmol HCl}} = 2.485 \text{ mmol Na}_2\text{CO}_3$$

Since Na_2CO_3 is the limiting reactant,

(a)
$$1.057 \text{ mmol} \times \frac{1 \text{ mmol CO}_2}{1 \text{ mmol Na}_2\text{CO}_3} \times \frac{44.01 \text{ g}}{\text{mol}} \times \frac{1 \text{ mol}}{1000 \text{ mmol}} = \underline{\underline{0.04652 \text{ g CO}_2}}$$

(b)
$$\frac{4.97 \text{ mmol} - 2 \times 1.057 \text{ mmol}}{100.0 \text{ mL}} = \underline{\underline{0.0286 \text{ M in HCl}}}$$

3-35. $Na_2SO_3 + 2HClO_4 \rightarrow 2NaClO_4 + H_2O + SO_2$

$$75.00\ \text{mL} \times 0.3333\ \text{M} = 25.00\ \text{mmol}\ Na_2SO_3$$

$$150.0\ \text{mL} \times 0.3912\ \text{M} = 58.68\ \text{mmol}\ HClO_4$$

$$25.00\ \text{mmol}\ Na_2SO_3 \times \frac{2\ \text{mmol}\ HClO_4}{1\ \text{mmol}\ Na_2SO_3} = 50.00\ \text{mmol}\ HClO_4$$

$$58.68\ \text{mmol}\ HClO_4 \times \frac{1\ \text{mmol}\ Na_2SO_3}{2\ \text{mmol}\ HClO_4} = 29.34\ \text{mmol}\ Na_2SO_3$$

Since Na_2SO_3 is the limiting reactant,

(a)

$$25.00\ \text{mmol}\ Na_2SO_3 \times \frac{1\ \text{mmol}\ SO_2}{1\ \text{mmol}\ Na_2SO_3} \times \frac{1\ \text{mol}}{1000\ \text{mmol}} \times \frac{64.06\ \text{g}\ SO_2}{1\ \text{mmol}\ SO_2} = \underline{\underline{1.602\ \text{g}\ SO_2}}$$

(b)

$$\frac{58.68\ \text{mmol} - 50.00\ \text{mmol}}{150.0\ \text{mL} + 75.00\ \text{mL}} = \underline{\underline{0.0386\ \text{M in}\ HClO_4}}$$

3-37.

$$\frac{26.43\ \text{g KI}}{10^3\ \text{g soln}} \times \frac{1000\ \text{g soln}}{\text{L soln}} \times \frac{1\ \text{mol KI}}{166.0\ \text{g KI}} = 0.1592\frac{\text{mol KI}}{\text{L}}$$

$$0.1592\frac{\text{mmol KI}}{\text{mL soln}} \times 200\ \text{mL} \times \frac{1\ \text{mmol}\ AgNO_3}{1\ \text{mmol KI}} \times \frac{1\ \text{ml}\ AgNO_3}{0.01000\ \text{mmol}\ AgNO_3} = \underline{\underline{3184\ \text{mL}\ AgNO_3}}$$

Chapter 4

4-1. **(a)** A weak electrolyte is a substance that ionizes only partially in a solvent.

(c) The conjugate base of a Brønsted-Lowry acid is the species formed when the acid has donated a proton.

(e) An amphiprotic solute is one that can act either as an acid or as a base when dissolved in a solvent.

(g) Autoprotolysis is self-ionization of a solvent to give a conjugate acid and a conjugate base.

(i) The Le Châtelier principle states that the position of equilibrium in a system always shifts in a direction that tends to relieve an applied stress to the system.

4-2. **(a)** An amphiprotic solute is a solute that acts as a base with acidic solutes and as an acid with basic solutes.

(c) A leveling solvent is one in which a series of acids (or bases) all dissociate completely.

4-3. For an aqueous equilibrium in which water is a reactant or a product, the concentration of water is normally so much larger than the concentrations of the reactants and products that its concentration can be assumed to be constant and independent of the position of the equilibrium. Thus its concentration is assumed to be constant and is included in the equilibrium constant. For a solid reactant or product, it is the concentration of that reactant in the solid phase that would influence the position of equilibrium. However, the concentration of a species in the solid phase is constant. Thus as long as some solid exists as a second phase, its effect on the equilibrium is constant, and its concentration is included in the equilibrium constant.

4-4.

	Acid	Conjugate Base
(a)	$HOCl$	OCl^-
(c)	NH_4^+	NH_3
(e)	$H_2PO_4^-$	HPO_4^{2-}

4-5.

	Base	Conjugate Acid
(a)	H_2O	H_3O^+
(c)	H_2O	H_3O^+
(e)	PO_4^{3-}	HPO_4^{2-}

4-6. **(a)** $2H_2O \rightleftarrows H_3O^+ + OH^-$

(c) $2CH_3NH_2 \rightleftarrows CH_3NH_3^+ + CH_3NH^-$

4-7. **(a)** $C_2H_5NH_2 + H_2O \rightleftarrows C_2H_5NH_3^+ + OH^-$

$$K_b = \frac{K_w}{K_a} = \frac{1.00 \times 10^{-14}}{2.31 \times 10^{-11}} = \frac{[C_2H_5NH_3^+][OH^-]}{[C_2H_5NH_2]} = 4.33 \times 10^{-4}$$

(c) $C_5H_5NH^+ + H_2O \rightleftarrows C_5H_5N + H_3O^+$

$$K_a = \frac{[H_3O^+][C_5H_5N]}{[C_5H_5NH^+]} = 5.90 \times 10^{-6}$$

(e) $H_3AsO_4 + 3H_2O \rightleftarrows 3H_3O^+ + AsO_4^{3-}$

$$K_1K_2K_3 = 5.8 \times 10^{-3} \times 1.1 \times 10^{-7} \times 3.2 \times 10^{-12}$$

$$2.0 \times 10^{-21} = \frac{[H_3O^+]^3[AsO_4^{3-}]}{[H_3AsO_4]} = \beta_3$$

4-8. **(a)** $AgIO_3(s) \rightleftarrows Ag^+ + IO_3^-$ $\quad K_{sp} = [Ag^+][IO_3^-]$

(b) $Ag_2SO_3(s) \rightleftarrows 2Ag^+ + SO_3^{2-}$ $\quad K_{sp} = [Ag^+]^2[SO_3^{2-}]$

(c) $Ag_3AsO_4(s) \rightleftarrows 3Ag^+ + AsO_4^{3-}$ $\quad K_{sp} = [Ag^+]^3[AsO_4^{3-}]$

(d) $PbClF(s) \rightleftarrows Pb^{2+} + Cl^- + F^-$ $\quad K_{sp} = [Pb^{2+}][Cl^-][F^-]$

4-9. **(a)** $S = [Ag^+] = [IO_3^-]$ $\quad K_{sp} = [Ag^+][IO_3^-] = \underline{\underline{S^2}}$

(b) $S = [SO_3^{2-}] = \frac{1}{2}[Ag^+]$ $\qquad K_{sp} = [Ag^+]^2[SO_3^{2-}] = (2S)^2(S) = \underline{\underline{4S^3}}$

(c) $S = [AsO_4^{3-}] = \frac{1}{3}[Ag^+]$ $\qquad K_{sp} = (3S)^3(S) = \underline{\underline{27S^4}}$

(d) $S = [Pb^{2+}][Cl^-][F^-]$ $\qquad K_{sp} = S \times S \times S = \underline{\underline{S^3}}$

4-10. (a) $[Ag^+] = [VO_3^-] = 7.1 \times 10^{-4}\ \text{mL/L}$

$K_{sp} = (7.1 \times 10^{-4})(7.1 \times 10^{-4}) = \underline{\underline{5.0 \times 10^{-7}}}$

(c) $[Pb^{2+}] = 4.3 \times 10^{-5}$ $\qquad [IO_3^-] = 2 \times 4.3 \times 10^{-5} = 8.6 \times 10^{-5}$

$K_{sp} = (4.3 \times 10^{-5})(8.6 \times 10^{-5})^2 = \underline{\underline{3.2 \times 10^{-13}}}$

(e) $[Th^{4+}] = 3.3 \times 10^{-4}$ $\qquad [OH^-] = 4 \times 3.3 \times 10^{-4} = 13.2 \times 10^{-4}$

$K_{sp} = (3.3 \times 10^{-4})(13.2 \times 10^{-4})^4 = \underline{\underline{1.0 \times 10^{-15}}}$

4-11. (a) $5.0 \times 10^{-7} = [Ag^+][VO_3^-] = (0.050)S$

$\underline{\underline{S = 1.0 \times 10^{-5}\ \text{mol/L}}}$

(c) $3.2 \times 10^{-13} = [Pb^{2+}][IO_3^-]^2 = (0.05)[IO_3^-]^2$

$[IO_3^-] = 2.53 \times 10^{-6}$ $\qquad S = \frac{1}{2}[IO_3^-] = \underline{\underline{1.3 \times 10^{-6}\ \text{mol/mL}}}$

(e) $1.0 \times 10^{-15} = [Th^{4+}][OH^-]^4 = (0.050)(S/4)^4$

$$S/4 = \left(\frac{1.0 \times 10^{-15}}{0.050}\right)^{1/4} = 3.8 \times 10^{-4}\ \text{mol/L}$$

$$S = 1.5 \times 10^{-3}\ \text{mol/L}$$

4-12. **(a)** $5 \times 10^{-7} = [Ag^{+}][IO_3^{-}] = [Ag^{+}](0.05)$

$$[Ag^{+}] = S = 5 \times 10^{-7}/0.050 = 1.0 \times 10^{-5}\ \text{M}$$

(c) $3.2 \times 10^{-13} = [Pb^{2+}][IO_3^{-}]^2 = [Pb^{2+}](0.050)^2$

$$[Pb^{2+}] = S = 3.2 \times 10^{-13}/(0.050)^2 = 1.3 \times 10^{-10}\ \text{M}$$

(e) $1.0 \times 10^{-15} = [Th^{4+}](0.050)^4$

$$[Th^{4+}] = S = 1.0 \times 10^{-15}/(0.050)^4 = 1.6 \times 10^{-10}\ \text{M}$$

4-13. **(a)** $[Ag^{+}]^2[CrO_4^{2-}] = 1.2 \times 10^{-12}$

$$[CrO_4^{2-}] = 1.2 \times 10^{-12}/(2.12 \times 10^{-3})^2 = 2.7 \times 10^{-7}\ \text{M}$$

(b)
$$[CrO_4^{2-}] = \frac{1.2 \times 10^{-12}}{(1.00 \times 10^{-6})^2} = 1.2\ \text{M}$$

4-15. **(a)**
$$\frac{50.0\ \text{mL} \times 0.0500\ \text{M}\ Ce^{3+}}{50.00\ \text{mL} + 50.0\ \text{mL}} = 0.0250\ \text{M}\ Ce^{3+}$$

(b) Since the Ce^{2+} is in excess, assume that essentially all of the IO_3^{-} reacts so that

$$\text{no. mmol } Ce^{3}+ = \text{initial no. mmol } Ce^{2+} - \frac{1}{3}\text{initial no. mmol } IO_3^{-}$$

$$= 50.0\ \text{mL} \times 0.0500\ \text{M} - \frac{50.00\ \text{mL} \times 0.0500\ \text{M}}{3} = 1.67\ \text{mmol}$$

$$c_{Ce^{3+}} = \frac{1.67\ \text{mmol}}{100.0\ \text{mL}} = 1.67 \times 10^{-2}\ \text{M}$$

$[Ce^{3+}] = 1.67 \times 10^{-2} + S \approx 1.67 \times 10^{-2}$ (where S = solubility)

In order to check this assumption, we write

$$[IO_3^-] = 3S$$

$$K_{sp} = [Ce^{3+}][IO_3^-]^3 = 1.67 \times 10^{-2}(3S)^3 = 3.2 \times 10^{-10}$$

$$S = \left(\frac{3.2 \times 10^{-10}}{27 \times 1.67 \times 10^{-2}}\right)^{1/3} = 8.9 \times 10^{-4}$$

which is $<< 1.67 \times 10^{-2}$

and $[Ce^{2+}] = 1.67 \times 10^{-2} = \underline{\underline{1.7 \times 10^{-2}\ M}}$

(c) no. mmol IO_3^- = 50.0 mL × 0.150 M = 7.50 mmol

no. mmol Ce^{2+} = 50.0 mL × 0.050 M = 2.50 mmol

Since these equations represent the stoichiometric ratio, the solubility is calculated from K_{sp}. Here,

$$[Ce^{3+}] = S \quad \text{and} \quad [IO_3^-] = 3S$$

$$K_{sp} = [Ce^{3+}][IO_3^-]^3 = S \cdot (3S)^3 = 27S^4 = 3.2 \times 10^{-10}$$

$$S^4 = 1.18 \times 10^{-11}$$

$$[Ce^{3+}] = S = (1.18 \times 10^{-11})^{1/4} = \underline{\underline{1.9 \times 10^{-3}\ M}}$$

(d) no. mmol IO_3^- = 50.0 mL × 0.300 M = 15.0 mmol

no. mmol Ce^{2+} = 2.50 mmol

Let S = solubility and

$$[IO_3^-] = \frac{15.0\text{ mmol} - 3(2.50\text{ mmol})}{100\text{ mL}} + 3S = 0.0750\text{ M} + 3S$$

$$[Ce^{3+}] = S$$

$$K_{sp} = S(0.0750 + 3S)^3 = 3.2 \times 10^{-10}$$

Assume $3S << 0.0750$,

$$S = 3.2 \times 10^{-10}/(0.0750)^3 = \underline{\underline{7.6 \times 10^{-7}\text{ M}}}$$

4-17. (a) $S_{TlI} = [Tl^+] = [I^-] = \sqrt{K_{sp}} = \sqrt{6.5 \times 10^{-8}} = \underline{\underline{2.5 \times 10^{-4}\text{ M}}}$

$$S_{AgI} = [Ag^+] = [I^-] = \sqrt{8.3 \times 10^{-17}} = \underline{\underline{9.1 \times 10^{-9}\text{ M}}}$$

$$K_{sp} = [Pb^+][I^-]^2 = S(2S)^2 = 4S^3$$

$$S_{PbI_2} = (K_{sp}/4)^{1/3} = (7.1 \times 10^{-9}/4)^{1/3} = \underline{\underline{1.2 \times 10^{-3}\text{ M}}}$$

$$K_{sp} = [Bi^{3+}][I^-]^3 = S(3S)^3 = 27S^4$$

$$S_{BiI_3} = (K_{sp}/27)^{1/4} = (8.1 \times 10^{-19}/27)^{1/4} = \underline{\underline{1.3 \times 10^{-5}\text{ M}}}$$

$\underline{\underline{PbI_2 > TlI > BiI_3 > AgI}}$

(b) $S_{TlI} = [Tl^+] = K_{sp}/[I^-] = 6.5 \times 10^{-8}/0.10 = \underline{\underline{6.5 \times 10^{-7}\text{ M}}}$

$$S_{AgI} = [Ag^+] = K_{sp}/[I^-] = 8.3 \times 10^{-17}/0.10 = \underline{\underline{8.3 \times 10^{-16}\text{ M}}}$$

$$S_{PbI_2} = [Pb^{2+}] = K_{sp}/[I]^2 = 7.1 \times 10^{-9}/(0.10)^2 = \underline{\underline{7.1 \times 10^{-7}\text{ M}}}$$

$$S_{BiI_3} = [BiI_3 = [Bi^{3+}] = K_{sp}/[I^-]^2 = 8.1 \times 10^{-19}/(0.10)^3 = \underline{\underline{8.1 \times 10^{-16}\text{ M}}}$$

$\underline{\underline{PbI_2 > TlI > AgI > BiI_3}}$

(c) $S_{\text{TlI}} = [\text{I}^-] = K_{sp}/[\text{Tl}^+] = 6.5\times10^{-8}/0.010 = \underline{\underline{6.5\times10^{-6}\text{ M}}}$

$S_{\text{AgI}} = [\text{I}^-] = K_{sp}/[\text{Ag}^+] = 8.3\times10^{-17}/0.010 = \underline{\underline{8.3\times10^{-15}\text{ M}}}$

$S_{\text{PbI}_2} = [\text{I}^-]/2 \qquad [\text{I}^-] = \sqrt{K_{sp}/[\text{Pb}^{2+}]}$

$S_{\text{PbI}_2} = \dfrac{1}{2}\sqrt{K_{sp}/[\text{Pb}^{2+}]} = \dfrac{1}{2}\sqrt{7.1\times10^{-9}/0.010} = \underline{\underline{4.2\times10^{-4}\text{ M}}}$

$S_{\text{BiI}_3} = [\text{I}^-]/3 \qquad [\text{I}^-] = (K_{sp}/[\text{Bi}^{3+}])^{1/3}$

$S_{\text{BiI}_3} = \dfrac{1}{3}(K_{sp}/[\text{Bi}^{3+}])^{1/3} = \dfrac{1}{3}(8.1\times10^{-19}/0.010)^{1/3} = \underline{\underline{1.4\times10^{-6}\text{ M}}}$

$\underline{\underline{\text{PbI}_2 > \text{TlI} > \text{BiI}_3 > \text{AgI}}}$

4-20. (a) $\dfrac{[\text{H}_3\text{O}^+][\text{OCl}^-]}{[\text{HOCl}]} = 3.0\times10^{-8} = K_a$

$[\text{H}_3\text{O}^+] = [\text{OCl}^-]$ and $[\text{HOCl}] = 0.0200 - [\text{H}_3\text{O}^+]$

Assume $[\text{H}_3\text{O}^+] << 0.0200$. Then

$[\text{H}_3\text{O}^+]^2/0.0200 = 3.0\times10^{-8}$

$[\text{H}_3\text{O}^+] = \sqrt{0.0200\times3.0\times10^{-8}} = \underline{\underline{2.4\times10^{-5}\text{ M}}}$

The assumption appears valid and

$[\text{OH}^-] = 1.00\times10^{-14}/(2.4\times10^{-5}) = \underline{\underline{4.1\times10^{-10}\text{ M}}}$

(c) $\text{C}_2\text{H}_5\text{NH}_2 + \text{H}_2\text{O} \rightleftarrows \text{C}_2\text{H}_5\text{NH}_3^+ + \text{OH}^- \qquad K_b = \dfrac{K_w}{K_a} = \dfrac{1.00\times10^{-14}}{2.31\times10^{-11}}$

$= 4.33\times10^{-4}$

$[\text{OH}^-] = [\text{C}_2\text{H}_5\text{NH}_3^+]$

$[C_2H_5NH_2] = 0.200 - [OH^-] \approx 0.200$

$$\frac{[OH^-]^2}{0.200} = 4.33 \times 10^{-4}$$

$[OH^-] = \sqrt{4.33 \times 10^{-4} \times 0.200} = \underline{\underline{9.3 \times 10^{-3}\ M}}$

$[H_3O^+] = 1.00 \times 10^{-14} / 9.3 \times 10^{-3} = \underline{\underline{1.07 \times 10^{-12}}}$

(e) $OCl^- + H_2O \rightleftarrows HOCl + OH^-$

$$\frac{[HOCl][OH^-]}{[OCl^-]} = K_b = \frac{K_w}{K_a} = \frac{1.00 \times 10^{-14}}{3.0 \times 10^{-8}} = 3.33 \times 10^{-7}$$

$[OH^-] = [HOCl]$ and $[OCl^-] = 0.120 - [OH^-] \approx 0.120$

$[OH^-]^2/0.120 = 3.3 \times 10^{-7}$

$[OH^-] = \sqrt{0.120 \times 3.33 \times 10^{-7}} = \underline{\underline{2.0 \times 10^{-4}\ M}}$

The assumption is valid, and

$[H_3O^+] = 1.00 \times 10^{-14} / (2.00 \times 10^{-4}) = \underline{\underline{5.0 \times 10^{-11}\ M}}$

(g) $HONH_3^+ + H_2O \rightleftarrows HONH_2 + H_3O^+ \quad K_a = 1.10 \times 10^{-6}$

$[H_3O^+] = [HONH_2]$

$[HONH_3^+] = 0.100 - [H_3O^+] \approx 0.100$

$[H_3O^+]^2 = 0.100 \times 1.10 \times 10^{-6}$

$[H_3O^+] = \underline{\underline{3.32 \times 10^{-4}\ M}}$ and $[OH^-] = 1.00 \times 10^{-14} / 3.32 \times 10^{-4}$

$= \underline{\underline{3.02 \times 10^{-11}\ M}}$

4-22. **(a)** $HA + H_2O \rightleftarrows H_3O^+$

$$K_a = \frac{[H_2O^+][A^-]}{[HA]} = \frac{[H_3O^+]}{c_{HA} - [H_3O^+]} = 1.36 \times 10^{-3}$$

Because K_a is relatively large, we must solve the quadratic equation. We do this by successive approximations.

$$[H_3O^+] = \sqrt{K_a(c_{HA} - [H_3O^+])}$$

$$[H_3O^+] = \sqrt{1.36 \times 10^{-3}(0.100 - 0)} = 1.17 \times 10^{-2}$$

$$[H_3O^+] = \sqrt{1.36 \times 10^{-3}(0.100 - 1.17 \times 10^{-2})} = 1.096 \times 10^{-2}$$

$$= \underline{\underline{1.10 \times 10^{-2}\ M}} \qquad \text{(good to 3 significant figures)}$$

(c)

$$CH_3NH_2 + H_2O \rightarrow CH_3NH_3^+ + OH^- \qquad K_b = \frac{1.00 \times 10^{-14}}{2.3 \times 10^{-11}} = 4.35 \times 10^{-4}$$

$$[OH^-] = [CH_3NH_3^+] \text{ and } [cH_3NH_2] = 0.0100 - [OH^-] \approx 0.0100$$

$$\frac{[OH^-]^2}{0.0100} = 4.35 \times 10^{-4}$$

$$[OH^-] = \sqrt{4.35 \times 10^{-6}} = 2.1 \times 10^{-3}\ M$$

We see, however, that the approximation is not very good. Thus, we write

$$\frac{[OH^-]^2}{0.0100 - [OH^-]} = 4.38 \times 10^{-4}$$

$$[OH^-]^2 + 4.38 \times 10^{-4}[OH^-] - 4.38 \times 10^{-6} = 0$$

Solving the quadratic gives $[OH^-] = 1.88 \times 10^{-3}$

$$[H_3O^+] = 1.00 \times 10^{-14}/(1.88 \times 10^{-3}) = \underline{\underline{5.3 \times 10^{-12}\ M}}$$

(e) $C_6H_5NH_3^+ + H_2O \rightleftarrows C_6H_5NH_2 + H_3O^+ \qquad K_a = 2.51 \times 10^{-5}$

Proceeding as in part (c), we obtain after solving the quadratic expressions, $[H_3O^+] = \underline{\underline{1.46 \times 10^{-4}}}$. (Using the simplifying assumption, we find $[H_3O^+] = 1.58 \times 10^{-4}$.)

Chapter 5

5-1. **(a)** *Constant errors* are the same magnitude regardless of the sample size. *Proportional errors* are proportional in size to the sample size.

(c) The *mean* is the sum of the measurements in a set divided by the number of measurements. The *median* is the central value for a set of data; half of the measurements are larger and half are smaller than the median.

5-2. (1) Random temperature fluctuations causing random changes in the length of the metal rule; (2) uncertainties arising from having to move and position the rule twice; (3) personal judgment in reading the rule; (4) vibrations in the table and/or rule; (5) uncertainty in locating the rule perpendicular to the edge of the table.

5-3. The three types of systematic error are *instrumental error, method error*, and *personal error.*

5-5. Constant errors.

5-6. **(a)** $\dfrac{-0.4\text{ mg}}{900\text{ mg}}\times 100\% = \underline{\underline{-0.04\%}}$

(c) $\dfrac{-0.4\text{ mg}}{150\text{ mg}}\times 100\% = \underline{\underline{-0.3\%}}$

5-7. **(a)** $\dfrac{0.4\text{ mg Au lost}}{\text{mg Au in sample}}\times 100\% = 0.2\%\text{ error}$

$$\frac{0.4\text{ mg Au}\times 100\%}{0.2\%} = 200\text{ mg Au in sample}$$

$$\frac{200\text{ mg Au in sample}}{\text{mg sample}}\times 100\% = 1.2\%\text{ Au}$$

$$\text{mg sample} = \frac{200\text{ mg Au}\times 100\%}{1.2} = 1.7\times 10^{4}\text{ mg sample}$$

$$= \underline{\underline{17\text{ g sample}}}$$

Proceeding in the same way, we obtain

(c) 4 g sample

5-8. **(a)** $\frac{0.04\text{ mL}}{50.00\text{ mL}} \times 100\% = 0.08\%$

Proceeding in a similar way, we obtain

(c) 0.16%

5-9. **(a)** $E_r = \frac{-0.4\text{ mg}}{40\text{ mg}} \times 100\% = -1.0\%$

Proceeding in the same way, we obtain

(c) –0.10%

5-10. **(a)**

x_i	$d_i = \lvert x_i - \bar{x} \rvert$
0.0110	0.0004
0.0104	0.0002
0.0105	0.0001
Σx_i = 0.0319	0.0007

$\bar{x} = 0.0319/3 = 0.0106 = \text{mean}$

median = 0.0105

$\bar{d} = 0.0007/3 = 0.0002 = \text{mean deviation}$

Proceeding in the same way, we obtain

	Mean	Median	Deviation from Mean	Mean Deviation
(a)	0.0106	0.0105	0.0004, 0.0002, 0.0001	0.0002
(c)	190	189	2, 0, 4, 3	2
(e)	39.59	39.64	0.24, 0.02, 0.34, 0.09	0.17

5-10. (a), (c)

	A	B	C	D	E	F	G	H
1	**Problem 5-10**		**(a)**				**(c)**	
2			***x***	***d***			***x***	***d***
3			0.01100	0.00037			188.00	1.75
4			0.01040	0.00023			190.00	0.25
5			0.01050	0.00013			194.00	4.25
6							187.00	2.75
7		**mean**	0.01063	0.00024		**mean**	189.75	2.25
8		**median**	0.01050			**median**	189.00	
9								
10	**Spreadsheet Documentation**						**(e)**	
11							***x***	***d***
12	**C7=AVERAGE(C3:C6)**						39.83	0.24
13	**C8=MEDIAN(C3:C6)**						39.61	0.02
14	**D3=ABS(C3-$C$7)**						39.25	0.34
15	**D7=AVERAGE(D3:D6)**						39.68	0.09
16						**mean**	39.59	0.17
17						**median**	39.65	

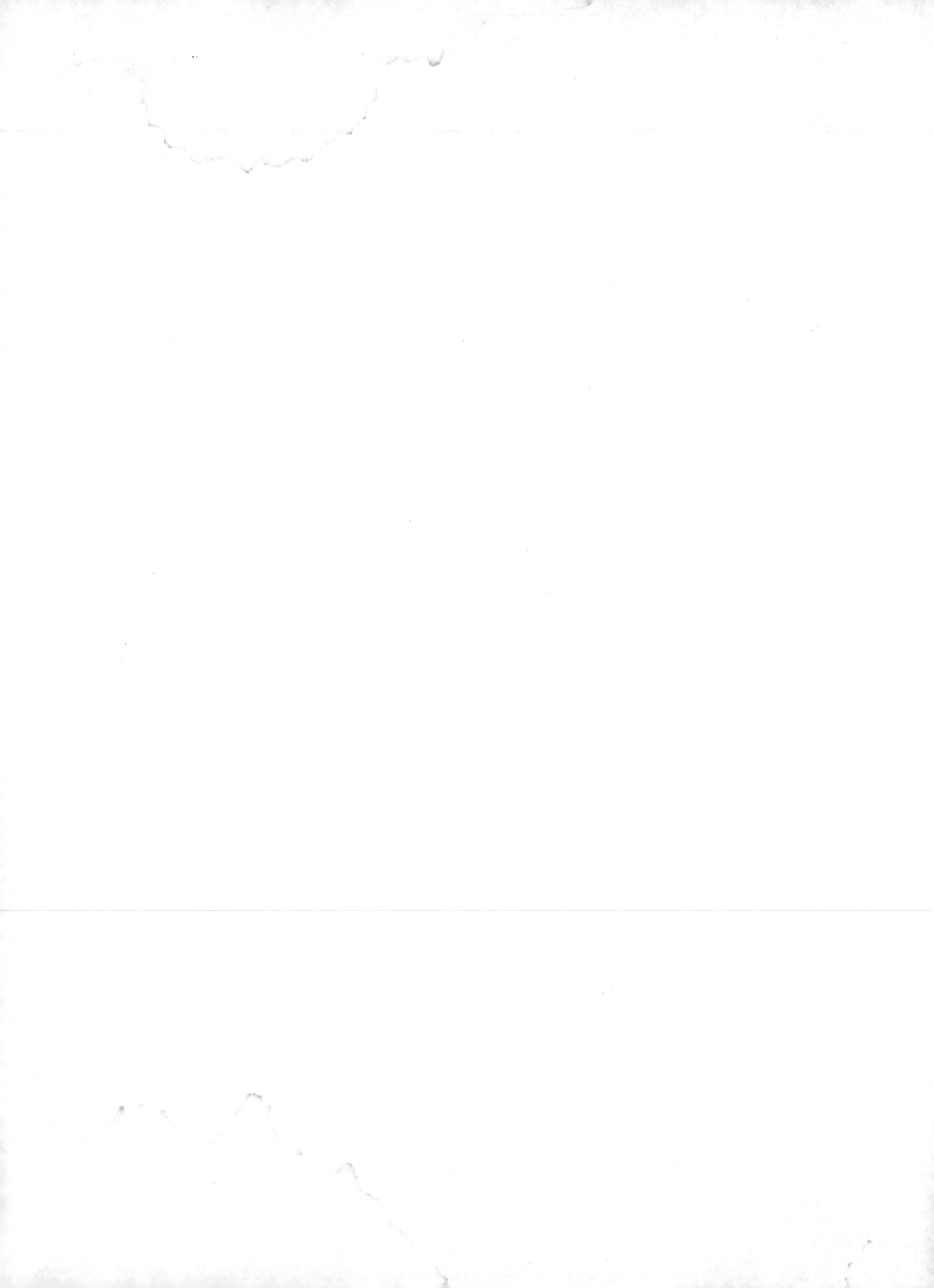

Chapter 6

6-1. **(a)** The *spread* or *range* for a set of replicate data is the numerical difference between the highest and lowest value.

(c) *Significant figures* include all the digits in a number that are known with certainty plus the first uncertain digit.

6-2. **(a)** The *sample variance*, s^2, is given by the expression

$$s^2 = \frac{\sum_{i=1}^{N} (x_i - \bar{x})^2}{N-1}$$

where $\bar{x}$ is the sample mean.

The *sample standard deviation* is given by

$$s = \sqrt{\frac{\sum_{i=1}^{N} (x_i - \bar{x})^2}{N-1}}$$

(c) *Accuracy* represents the agreement between an experimentally measured value and the true or accepted value. *Precision* describes the agreement among measurements that have been performed in exactly the same way.

6-3. **(a)** In statistics, a sample is a small set of replicate measurement. In chemistry, a sample is a portion of a material that is used for analysis.

6-5. For A:

x_i	x_i^2
3.5	12.25
3.1	9.61
3.1	9.61
3.3	10.89
2.5	6.25
$\Sigma x_i = 15.5$	$\Sigma x_i^2 = 48.61$

(a) $\bar{x} = 15.5/5 = 3.1$

(b) median = 3.1

(c) spread = w = 3.5 - 2.5 = 1.0

(d)

$$s = \sqrt{\frac{\Sigma x_i^2 - (\Sigma x_i)^2/N}{N-1}} = \sqrt{\frac{48.61 - (15-5)^2/5}{5-1}} = 0.37 = 0.4$$

(e) CV $= (s/\bar{x}) \times 100\% = (0.37/3.1) \times 100\% = 12\%$

Results for sets A, C, and E obtained in a similar way, are given in the following table.

	A	C	E
N	5	4	4
Σx_i	15.5	3.298	282.13
Σx_i^2	48.61	2.727	19,899
(a) $\bar{x}$	3.1	0.824	70.53
(b) median	3.1	0.803	70.64
(c) w	1.0	0.108	0.44
(d) s	0.37	0.051	0.22
(e) CV	12%	6.2%	0.30%

6-6. For Set A, $E = 3.1 - 3.0 = 0.1$

$$E_r = (0.1/3.0) \times 1000 \text{ ppt} = 33 \text{ ppt}$$

Proceeding in the same way, we obtain

Set C - 0.006 and -7 ppt

Set E 0.48 and 6.9 ppt

6-7. **(a)** $s_y^2 = (0.03)^2 + (0.001)^2 + (0.001)^2 = 9.02 \times 10^{-4}$

$$s_y = \sqrt{9.02 \times 10^{-4}} = \underline{\underline{0.03}}$$

$$\text{CV} = \frac{0.03}{-1.438} \times 100\% = \underline{\underline{-2\%}}$$

$$y = \underline{\underline{-1.44\,(\pm 0.03)}}$$

Note that the answer should be -1.438 and not +0.438.

(c)

$$\frac{s_y}{y} = \sqrt{\left(\frac{0.3}{66.2}\right)^2 + \left(\frac{0.02\times10^{-17}}{1.13\times10^{-17}}\right)^2} = 1.83\times10^{-2}$$

$$\text{CV} = \frac{s_y}{y}\times100\% = \underline{\underline{1.8\%}}$$

$$s_y = 1.83\times10^{-2}\times7.48\times10^{-16} = \underline{\underline{0.14\times10^{-16}}}$$

$$y = \underline{\underline{7.5\,(\pm0.1)\times10^{-16}}}$$

(e)

$$s_{\text{num}} = \sqrt{(6)^2+(3)^2} = 6.71$$

$$s_{\text{dem}} = \sqrt{(1)^2+(8)^2} = 8.06$$

$$y = \frac{157-59}{1220+77} = \frac{98.0}{1297} = 0.0755$$

$$\frac{s_y}{y} = \sqrt{\left(\frac{6.71}{98.0}\right)^2 = \left(\frac{8.06}{1291}\right)^2} = 6.88\times10^{-2} = 6.9\times10^{-2}$$

$$\text{CV} = 6.9\times10^{-2}\times100\% = \underline{\underline{6.9\%}}$$

$$s_y = 6.88\times10^{-2}\times0.0755 = 5.19\times10^{-3} = \underline{\underline{0.5\times10^{-2}}}$$

$$y = \underline{\underline{7.6\,(\pm0.5)\times10^{-2}}}$$

6-8. (a)

$$s_y = \sqrt{(0.02\times10^{-8})^2+(0.2\times10^{-9})^2} = 2.83\times10^{-10} = \underline{\underline{0.03\times10^{-8}}}$$

$$\text{CV} = \frac{2.83\times10^{-10}}{1.374\times10^{-9}}\times100\% = 2.06\% = \underline{\underline{2\%}}$$

$$y = \underline{\underline{-1.37\,(\pm0.03)\times10^{-8}}}$$

(c)

$$\frac{s_y}{y} = \sqrt{\left(\frac{0.0005}{0.002}\right)^2+\left(\frac{0.02}{20.2}\right)^2+\left(\frac{1}{300}\right)^2} = 0.250$$

$$\text{CV} = \underline{\underline{25\%}}$$

$$s_y = 0.250 \times 12.12 = \underline{\underline{3}}$$

$$y = \underline{\underline{12 \pm 3}}$$

(e)

$$\frac{s_y}{y} = \sqrt{\left(\frac{1}{100}\right)^2 + \left(\frac{1}{2}\right)^2} = 0.50$$

$$\text{CV} = 0.50 \times 100\% = \underline{\underline{50\%}}$$

$$s_y = 50 \times 0.50 = \underline{\underline{25}}$$

$$y = \underline{\underline{50\,(\pm 25)}}$$

6-9. **(a)** For Set 1

$$s_1 = \sqrt{\frac{\sum_{i=1}^{N} (x_1 - \bar{x})}{N-1}} = \sqrt{\frac{(0.13)^2 + (0.09)^2 + (0.08)^2 + (0.06)^2 + (0.08)^2}{5-1}}$$

$$= \sqrt{\frac{0.04140}{4}} = \underline{\underline{0.10}}$$

The results for all of the sets follow.

Set	$\Sigma(x_i - x)^2$	s_i, % K^+	N
1	0.0414	0.10	5
2	0.0289	0.12	3
3	0.0462	0.12	4
4	0.0326	0.10	4
5	0.0460	0.11	5
Total	0.1951		21

$$s_{\text{pooled}} = 0.11\%\ \text{K}$$

6-9.

	A	B	C	D	E	F	G	H
1	**Problem 6-9**							
2								
3	**Sample**	**1**	**2**	**3**	**4**	**5**	**No. Sets**	
4		0.13	0.09	0.02	0.12	0.08	5	
5		0.09	0.08	0.17	0.06	0.06		
6		0.08	0.12	0.05	0.05	0.14		
7		0.06		0.12	0.11	0.10		
8		0.08				0.08	**Total**	
9	**N**	5	3	4	4	5	21	
10	**Sum of Squares**	0.041	0.029	0.046	0.033	0.046	0.195	
11	**s**	0.102	0.120	0.124	0.104	0.107	0.110	**s(pool)**
12								
13	**Spreadsheet Documentation**							
14								
15	**B9=COUNT(B4:B8)**							
16	**B10=B4^2+B5^2+B6^2+B7^2+B8^2**							
17	**B11=SQRT(B10/(B9-1))**							
18	**G4=COUNT(B4:F4)**							
19	**G9=SUM(B9:F9)**							
20	**G10=SUM(B10:F10)**							
21	**G11=SQRT(G10/(G9-G4))**							

6-11.

Sample	$\bar{x}$	$\Sigma(x_i - \bar{x})^2$	Sample	$\bar{x}$	$\Sigma(x_i - \bar{x})^2$
1	2.255	4.50×10^{-4}	6	1.045	1.25×10^{-3}
2	8.55	4.50×10^{-2}	7	14.60	8.00×10^{-2}
3	7.55	5.00×10^{-3}	8	21.50	3.20×10^{-1}
4	12.25	2.45×10^{-1}	9	8.60	8.00×10^{-2}
5	4.25	5.00×10^{-3}			$\Sigma\Sigma$ 0.7817

$$s_{\text{pooled}} = \sqrt{\frac{0.7817}{18-9}} = \underline{\underline{0.29\% \text{ heroin}}}$$

	A	B	C	D	E	F	G	H
1	**Problem 6-11**							
2								
3	**Sample**	**x1**	**x2**	**s^2**	**N**	**DF**	**Dev^2**	
4	1	2.24	2.27	0.00045	2	1	0.00045	
5	2	8.40	8.70	0.045	2	1	0.045	
6	3	7.60	7.50	0.005	2	1	0.005	
7	4	11.90	12.60	0.245	2	1	0.245	
8	5	4.30	4.20	0.005	2	1	0.005	
9	6	1.07	1.02	0.00125	2	1	0.00125	
10	7	14.40	14.80	0.08	2	1	0.08	
11	8	21.9	21.1	0.32	2	1	0.32	
12	9	8.8	8.4	0.08	2	1	0.08	
13					9	9	0.7817	0.29
14					**No. Sets**	**Tot. DF**	**Tot. Dev^2**	**s(pool)**
15								
16	**Spreadsheet Documentation**							
17								
18	**D4=VAR(B4:C4)**							
19	**E4=COUNT(B4:C4)**							
20	**F4=E4-1**							
21	**G4=D4*F4**							
22	**E13=COUNT(E4:E12)**							
23	**F13=SUM(F4:F12)**							
24	**G13=SUM(G4:G12)**							
25	**H13=SQRT(G13/F13)**							

Chapter 7

7-1. For Set A:

x_i	x_i^2
2.4	5.76
2.1	4.41
2.1	4.41
2.3	5.29
1.5	2.25
$\sum x_i = 10.4$	$\sum x_i^2 = 22.12$

$$\bar{x} = \frac{10.4}{5} = 2.08 = 2.1$$

$$s = \sqrt{\frac{22.12 - (10.4)^2/5}{4}} = 0.3493 = 0.35$$

$$95\%\ \text{CL} = 2.08 \pm \frac{2.78 \times 0.35}{\sqrt{5}} = 2.1 \pm 0.4$$

Results for Sets A, C, and E obtained in the same way.

	A	C	E
N	5	4	4
$\bar{x}$	2.08	0.0918	69.53
s	0.35	0.0055	0.22
95% CL	2.1 ± 0.4	0.092 ± 0.009	69.5 ± 0.03

7-2. $$CL = \bar{x} \pm \frac{z\sigma}{\sqrt{N}} \quad \text{Equation } 7-2$$

For Set A:

$$CL = 2.08 \pm \frac{1.96 \times 0.20}{\sqrt{5}} = 2.08\ (\pm 0.175) \quad \text{or} \quad \underline{\underline{2.1 \pm 0.2}}$$

Similarly, for the remaining sets:

For Set C:

$$CL = 0.0918 \pm \frac{1.96 \times 0.0070}{\sqrt{4}} = 0.0918 \pm 0.0069 \quad \text{or} \quad \underline{\underline{0.092 \pm 0.007}}$$

For Set E:

$$CL = 69.53 \pm \frac{1.96 \times 0.15}{\sqrt{4}} = 69.53 \pm 0.15 \quad \text{or} \quad \underline{\underline{69.5 \pm 0.2}}$$

The 95% confidence limit is an interval around a sample mean within which the population mean is expected to lie with 95% probability.

7-3. $$Q_{expt} = \frac{|x_q - x_n|}{w}$$

For Set A: $$Q_{expt} = \frac{|1.5 - 2.1|}{2.4 - 1.5} = 0.67$$

At 95% confidence $Q_{expt} = 0.71$ (see Table 7-3)

Thus the outlier *cannot be rejected* by the Q test.

Set	Q_{expt}	Q_{crit}	Result
A	0.67	0.710	retain
C	0.84	0.829	reject
E	0.95	0.829	reject

7-4. $\text{CL} = \bar{x} \pm z\sigma/\sqrt{N}$ (Equation 7 – 2)

(a)

$$80\%\ \text{CL} = 18.5 \pm \frac{(1.29)(2.4)}{\sqrt{1}} = 18.5 \pm 3.1\ \mu\text{g/mL} \quad \text{or} \quad \underline{\underline{18 \pm 3\ \mu\text{g/mL}}}$$

$$95\%\ \text{CL} = 18.5 \pm \frac{(1.96)(2.4)}{\sqrt{1}} = 18.5 \pm 4.7\ \mu\text{g/mL} \quad \text{or} \quad \underline{\underline{18 \pm 5\ \mu\text{g/mL}}}$$

(b)

$$80\%\ \text{CL} = 18.5 \pm \frac{(1.29)2.4}{\sqrt{2}} = 18.5 \pm 2.2\ \mu\text{g/mL} \quad \text{or} \quad 18 \pm 2\ \mu\text{g/mL}$$

$$95\%\ \text{CL} = 18.5 \pm \frac{(1.96)2.4}{\sqrt{2}} = 18.5 \pm 3.3\ \mu\text{g/mL} \quad \text{or} \quad \underline{\underline{18 \pm 3\ \mu\text{g/mL}}}$$

(c)

$$80\%\ \text{CL} = 18.5 \pm \frac{(1.29)(2.4)}{\sqrt{4}} = 18.5 \pm 1.5\ \mu\text{g/mL} \quad \text{or} \quad \underline{\underline{18 \pm 2\ \mu\text{g/mL}}}$$

$$95\%\ \text{CL} = 18.5 \pm \frac{(1.96)(2.4)}{\sqrt{4}} = 18.5 \pm 2.4\ \mu\text{g/mL} \quad \text{or} \quad \underline{\underline{18 \pm 2\ \mu\text{g/mL}}}$$

7-6. $z\sigma/\sqrt{N} = \mu\text{g/mL} \qquad \sqrt{N} = z\sigma/(1.5\ \mu\text{g/mL})$ (Equation 7 – 2)

95% $\quad \sqrt{N} = (1.96)(2.4)/1.5 = 3.14$

$N = (3.14)^2 = 9.83 = \underline{\underline{10\ \text{measurements}}}$

99% $\quad \sqrt{N} = (2.58)(2.4)/1.5 = 4.13$

$N = (4.13)^2 = 17.0 = \underline{\underline{17\ \text{measurements}}}$

7-8. **(a)**

x_i, meq/L	x_i^2
3.15	9.9225
3.25	10.5625
3.26	10.6276
$\Sigma x_i = 9.66$	$\Sigma x_i^2 = 31.1126$

$\bar{x} = 9.66/3 = 3.22$

$$s = \sqrt{\frac{\Sigma x_i^2 - (\Sigma x_i)^2/N}{N-1}} = \sqrt{\frac{31.1126 - (9.66)^2/3}{2}} = \sqrt{0.00370} = 0.061 \text{ mmol/L}$$

Substituing into Equation 7-4 gives

$$95\% \text{ CL} = \bar{x} \pm ts/\sqrt{N} = 3.22 \pm (4.30)(0.061)/\sqrt{3} = \underline{\underline{3.22 \pm 0.15 \text{ mmol/L}}}$$

(b) $$95\% \text{ CL} = 3.22 \pm (1.96)(0.056)/\sqrt{3} = \bar{x} \pm z\sigma/\sqrt{N}$$

$$= 3.22 \pm 0.063 \text{ mmol/L} \quad \text{or} \quad \underline{\underline{3.22 \pm 0.06 \text{ mmol/L}}}$$

7-10. **(a)** For 99% confidence, $z\sigma/\sqrt{N} = (2.58)(0.40)/\sqrt{N} = 0.3 \text{ mg/dL}$

$$\sqrt{N} = (2.58)(0.40)/0.3 = 3.44$$

$$N = (3.44)^2 = 11.8 \quad \text{or} \quad \underline{\underline{12 \text{ measurements}}}$$

7-11. **(a)** $(\bar{x} - \mu)_{\text{actual}} = 30.26 - 30.15 = 0.11$

$$(\bar{x} - \mu)_{\text{exp}} = zs_{\text{pooled}}/\sqrt{N} = (1.96)(0.094)/\sqrt{4} = 0.092 < 0.11$$

<u>Systematic error is indicated at 95% confidence</u>.

(b) $$(\bar{x} - \mu)_{\text{exp}} = ts_{\text{exp}}/\sqrt{N} = (3.18)(0.085)/\sqrt{4} = 0.135 > 0.11$$

<u>No systematic error is demonstrated</u>.

7-13.
$(\bar{x}_1 - \bar{x}_2)_{\text{actual}}$ is compared with $(\bar{x}_1 - \bar{x}_2)_{\text{calcu}} = \pm\sigma\sqrt{\dfrac{N_1 + N_2}{N_1 N_2}}$

where N_1 and $N_2 = 3$ and $z_{99\%} = 2.58$.

For As, $(\bar{x}_1 - \bar{x}_2)_{\text{actual}} \quad = \quad 129 - 119 \quad = \quad 10 \text{ ppm}$

$$(\bar{x}_1 - \bar{x}_2)_{\text{calc}} \quad = \quad \pm(2.58)(9.5)\sqrt{\frac{3+3}{3\times 3}} \quad = \quad \pm 20 \text{ ppm}$$

Results from similar calculations for the elements are listed the table that follows.

	Concentration, ppm		Difference, ppm			Difference at
Element	Clothes	Window	Actual	Calc	σ	99% CL
As	1.29	119	+ 10	± 20	9.5	no
Co	0.53	0.60	- 0.07	± 0.053	0.025	yes
La	3.92	3.52	+ 0.40	± 0.42	0.20	no
Sb	2.75	2.71	+ 0.04	± 0.53	0.25	no
Th	0.61	0.73	- 0.12	± 0.090	0.043	yes

7-15. Let $x_{i\text{C}}$ be nitrogen result from compound i

$x_{i\text{A}}$ be nitrogen result from air sample i

$$\bar{x}_{\text{C}} = \Sigma\frac{x_{i\text{C}}}{N_{\text{C}}} = \frac{9.19733}{4} = 2.29933$$

$$\bar{x}_{\text{A}} = \Sigma\frac{x_{i\text{A}}}{N_{\text{A}}} = \frac{6.93192}{3} = 2.31064$$

$$s_{\text{pooled}} = \sqrt{\frac{\Sigma(x_{i\text{C}} - \bar{x}_{\text{C}})^2 + \Sigma(x_{i\text{A}} - \bar{x}_{\text{A}})^2}{N_{\text{C}} - N_{\text{A}} - 2}}$$

$$= \sqrt{\frac{2.35968\times 10^{-6} + 1.50660\times 10^{-6}}{4+3-2}} = 8.79326\times 10^{-4}$$

$$\bar{x}_A - \bar{x}_C = 2.31064 - 2.29933 = 0.01131$$

Substituting into Equation 7-8, we find

$$t = \frac{\bar{x}_A - \bar{x}_C}{s_{\text{pooled}}\sqrt{\frac{N_1+N_2}{N_1 N_2}}} = \frac{0.01131}{8.79326 \times 10^{-4}\sqrt{\frac{4+3}{6\times 3}}} = 16.8$$

Since 16.8 is much larger than nearly all the values for t in Table 7-2, the hypothesis that A and C are alike is highly improbable.

7-16. (a)

$$Q_{\text{expt}} = \frac{|x_q - x_n|}{w} = \frac{|41.27 - 41.61|}{41.84 - 41.27} = 0.60$$

At 95% $Q_{\text{crit}} = 0.829$ (Table 7-4) > 0.60

Therefore the <u>outlier is retained</u>.

(b)

$$Q_{\text{exp}} = \frac{|7.388 - 7.295|}{|7.388 - 7.284|} = 0.89$$

$$Q_{\text{crit}} = 0.829$$

Therefore the <u>outlier is rejected</u>.

7-19.

	A	B	C	D	E	F	G	H	I
1	Problem 7-19 (a), (b)								
2									
3		*x*	*y*	y_{calc}					
4		5.00	-53.8	-55.84	*m*	-29.74	92.86	*b*	
5		4.00	-27.7	-26.1	s_{slope}	0.67616566	2.242587791	$s_{intercept}$	
6		3.00	2.7	3.64	r^2	0.998451638	2.138223562	s_r	
7		2.00	31.9	33.38	*F*	1934.531059	3	DF	
8		1.00	65.1	63.12	ss(reg)	8844.676	13.716	ss(resid)	
9					*N*	5			
10									
11–23									
24									
25		(c)	*E*	20.3		(d)	M_1	2	
26			pCa_x	2.44			s_{pCa1}	0.060	
27			S_{xx}	10.00			%RSD	2.5	
28			*M*	1					
29			$ybar_c$	3.67			M_2	8	
30			*ybar*	3.6			s_{pCa2}	0.041	
31			s_{pCa}	0.079			%RSD	1.7	
32			%RSD	3.2					
33									
34	Spreadsheet Documentation								
35									
36	D4=$F$4*B4+$G$4								
37	F4:G8=LINEST(C4:C8,B4:B8,TRUE,TRUE)								
38	F9=G7+2								
39	D26=(D25-G4)/F4								
40	D27=(F9-1)*STDEV(B4:B8)^2								
41	D30=AVERAGE(C4:C8)								
42	D31=ABS((G6/F4)*SQRT((1/D28)+(1/F9)+((D29-D30)^2/(F4^2*D27))))								
43	D32=100*D31/D26								

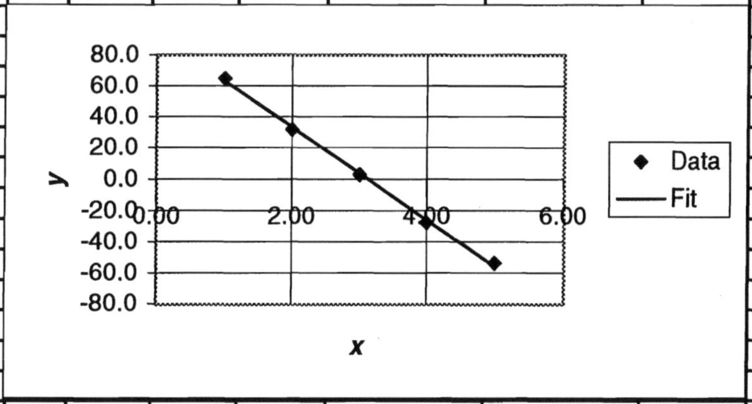

Chapter 8

8-1. **(a)** The individual particles of a *colloid* are smaller than about 10^{-5} mm in diameter, while those of a *crystalline precipitate* are larger. As a consequence, crystalline precipitates settle out of solution relatively rapidly, whereas colloidal particles do not unless they can be caused to agglomerate.

(c) *Precipitation* is the process by which a solid phase forms and is carried out of solution when the solubility product of a species is exceeded. *Coprecipitation* is the process in which a normally soluble species is carried out of solution during the formation of a precipitate.

(e) *Occlusion* is a type of coprecipitation in which an impurity is entrapped in a pocket formed by a rapidly growing crystal. *Mixed-crystal formation* is a type of coprecipitation in which a foreign ion is incorporated into a growing crystal in a lattice position that is ordinarily occupied by one of the ions of the precipitate.

8-2. **(a)** *Digestion* is a process for improving the purity and filterability of a precipitate by heating the solid in contact with the solution form which it is formed (the *mother liquor*).

(c) In *reprecipitation*, a precipitate is filtered, washed, redissolved, and then reformed from the new solution. Because the concentration of contaminant is lower in this new solution than in the original, the second precipitate contains less coprecipitated impurity.

(e) A *counter-ion layer* is a charged layer of solution that surrounds a colloidal particle.

(g) *Relative supersaturation* is given by the expression

$$\text{relative supersaturation} \quad = \quad \frac{Q - S}{S}$$

where Q is the concentration of a solute in a solution at any instant and S is its equilibrium solubility (Q - S).

8-3. A *chelating agent* is an organic compound that contains two or more electron-donor groups located in such a configuration that five- or six-membered rings are formed when the donor groups complex a cation.

8-5. **(a)** positive charge **(b)** adsorbed Ag^+ **(c)** NO_3^-

8-7. *Peptization* is the process in which a coagulated colloid returns to its original dispersed state as a consequence of a decrease in the electrolyte concentration of the solution in contact with the precipitate. Peptization of a coagulated colloid can be avoided by washing with an electrolyte solution rather than with pure water.

8-9. Throughout $\mathcal{M}$ stands for molar or atomic mass of a species.

(a)
$$\text{mass } SO_3 = \text{mass } BaSO_4 \times \frac{\mathcal{M}_{SO_3}}{\mathcal{M}_{BaSO_4}} = \text{mass } BaSO_4 \times \frac{80.06}{233.4}$$

(c)
$$\text{mass In} = \text{mass } In_2O_3 \times \frac{2 \times \mathcal{M}_{In}}{\mathcal{M}_{In_2O_3}} = \text{mass } In_2O_3 \times \frac{2 \times 114.8}{277.6}$$

(e)
$$\text{mass CuO} = \text{mass } Cu_2(SCN)_2 \times \frac{2 \times \mathcal{M}_{CuO}}{\mathcal{M}_{Cu_2(SCN)_2}} = \text{mass } Cu(SCN)_2 \times \frac{2 \times 79.54}{243.2}$$

(i)
$$\text{mass } Na_2B_4O_7 \cdot 10H_2O = \frac{2 \times \mathcal{M}_{B_2O_3}}{\mathcal{M}_{Na_2B_4O_7 \cdot 10H_2O}} = \frac{2 \times 69.62}{381.4}$$

8-10. $\mathcal{M}_{AgCl} = 143.32 \text{ g/mol} \qquad \mathcal{M}_{KCl} = 74.55 \text{ g/mol}$

$$\frac{0.7332 \text{ g AgCl} \times \frac{1 \text{molAgCl}}{143.32 \text{ g AgCl}} \times \frac{1 \text{ mol KCl}}{1 \text{ mol AgCl}} \times \frac{74.55 \text{ g KCl}}{\text{mol KCl}}}{0.4000 \text{ g sample}} \times 100\% = \underline{\underline{95.35\% \text{ KCl}}}$$

8-12. $\mathcal{M}_{Cu(IO_3)_2} = 413.35 \qquad \mathcal{M}_{CuSO_4 \cdot 5H_2O} = 249.69$

$$\text{mass } Cu(IO_3)_2 = 0.400 \text{ g} \times \frac{413.35}{249.69} = \underline{\underline{0.662 \text{ g}}}$$

8-14.
$$\text{mass AgI} = 0.240 \text{ g} \times \frac{30.6 \text{ g } MgI_2}{100 \text{ g}} \times \frac{1 \text{ } MgI_2}{278.11 \text{ g}} \times \frac{2 \text{ mol AgI}}{\text{mol } MgI_2} \times \frac{234.77 \text{ g AgI}}{\text{mol AgI}} = \underline{\underline{0.124 \text{ g}}}$$

8-16. $MgCO_3 + 2H^+ \rightarrow Mg^{2+} + CO_2 + H_2O$

$CO_2 + CaO \rightarrow CaCO_3$

$$\text{mass Mg} = 0.1881\text{ g CO}_2 \times \frac{1\text{ mol CO}_2}{44.01\text{ g}} \times \frac{1\text{ mol Mg}}{\text{mol CO}_2} \times \frac{24.305\text{ g Mg}}{\text{mol}} = 0.10388$$

$$\%\text{ Mg} = \frac{0.10388}{0.7406} \times 100\% = \underline{\underline{14.03}}$$

8-18. $\mathcal{M}_{BaCO_3} = 197.34$ g/mol

$$\frac{0.5613\text{ g BaCO}_3 \times \frac{1\text{ mol BaCO}_3}{197.34\text{ g BaCO}_3} \times \frac{1\text{ mol C}}{\text{mol BaCO}_3} \times \frac{12.011\text{ g C}}{\text{mol C}}}{0.1799\text{ g sample}} \times 100\% = \underline{\underline{18.99\%\text{ C}}}$$

8-20. $\mathcal{M}_{Hg_5(IO_6)_2} = 1448.8$ $\qquad$ $\mathcal{M}_{Hg_2Cl_2} = 472.09$

$$\frac{0.3408\text{ g Hg}_5(\text{IO}_6)_2 \times \frac{1\text{ mol Hg}_5(\text{IO}_6)_2}{1448.8\text{ g}} \times \frac{5\text{ mol Hg}_2\text{Cl}_2}{2\text{ mol Hg}_5(\text{IO}_6)_2} \times \frac{472.09\text{ g H}_2\text{Cl}_2}{\text{mol}}}{0.7152\text{ g sample}} \times 100\% = \underline{\underline{38.82\%}}$$

8-22. $\mathcal{M}_{NH_3} = 17.0306$ g/mol $\qquad$ $\mathcal{M}_{Pt} = 195.08$ g/mol

$$\frac{0.5881\text{ g Pt}}{0.2213\text{ g sample}} \times \frac{1\text{ mol Pt}}{195.08\text{ g}} \times \frac{2\text{ mol NH}_3}{1\text{ mol Pt}} \times \frac{17.0306\text{ g NH}_3}{\text{mol NH}_3} \times 100\% = \underline{\underline{46.40\%\text{ NH}_3}}$$

8-24. Let S_w = mass of sample in grams.

$\mathcal{M}_{BaSO_4} = 233.39$ g/mol $\qquad$ $\mathcal{M}_{SO_4^{-2}} = 96.064$ g/mol

$$0.300\text{ g BaSO}_4 \times \frac{1\text{ mol BaSO}_4}{233.39\text{ g BaSO}_4} \times \frac{1\text{ mol SO}_4^{2-}}{1\text{ mol BaSO}_4} = 1.2854 \times 10^{-3}\text{ mol SO}_4^{2-}$$

$$\frac{1.2854 \times 10^{-3}\text{ mol SO}_4^{2-} \times \frac{96.604\text{ g SO}_4^{2-}}{\text{mol}}}{S_w\text{ g sample}} \times 100\% = 20\%\text{ SO}_4^{2-}$$

$$S_w = \frac{0.12417\text{ g SO}_4^{2-} \times 100\%}{20\%\text{ SO}_4^{2-}} = \underline{\underline{0.621\text{ g sample}}}$$

We selected the smaller of the two percentages because it yields the smaller amount of $BaSO_4$. This quantity of sample yields at a level of 55% SO_4^{2-}:

$$0.621 \text{ g sample} \times \frac{55 \text{ g } SO_4^{2-}}{100 \text{ g sample}} \times \frac{1 \text{ mol } SO_4^{2-}}{96.064 \text{ g } SO_4^{2-}} \times \frac{1 \text{ mol } BaSO_4}{1 \text{ mol } SO_4^{2-}} \times \frac{233.39 \text{ g } BaSO_4}{\text{mol } BaSO_4} =$$

$$\underline{\underline{0.830 \text{ g } BaSO_4}}$$

8-26. **(a)** $\mathcal{M}_{AgCl} = 143.32 \text{ g/mol} \qquad \mathcal{M}_{ZrCl_4} = 233.03 \text{ g/mol}$

Let S_w = mass of sample in grams.

$$\frac{0.400 \text{ g AgCl} \times \frac{1 \text{ mol AgCl}}{143.32 \text{ g}} \times \frac{1 \text{mol } ZrCl_4}{4 \text{ mol AgCl}} \times \frac{233.03 \text{ g } ZrCl_4}{\text{mol } ZrCl_4}}{S_w \text{ g sample}} \times 100\% = 68\%$$

$$\frac{0.400 \text{ g AgCl} \times 0.40649 \text{ g } ZrCl_4/\text{g AgCl}}{S_w \text{ g sample}} \times 100\% = 68\%$$

$$S_w = 0.1626 \times 100/68 = \underline{\underline{0.239 \text{ g sample}}}$$

(b)

$$0.239 \text{ g sample} \times \frac{84 \text{ g } ZrCl_4}{100 \text{ g sample}} \times \frac{1 \text{ mol } ZrCl_4}{233.03 \text{ g } ZrCl_4} \times \frac{4 \text{ mol AgCl}}{1 \text{ mol } ZrCl_4} \times \frac{143.32 \text{ g AgCl}}{\text{mol AgCl}} =$$

$$\underline{\underline{0.494 \text{ g AgCl}}}$$

(c) The percent $ZrCl_4$ is given by

$$\% \, ZrCl_4 = 100 \text{ mass AgCl} = \text{mass AgCl} \times \frac{\frac{233.03}{143.32 \times 4}}{\text{mass sample}} \times 100$$

$$\text{mass sample} = \frac{\cancel{\text{mass AgCl}} \times 0.40649 \times \cancel{100}}{\cancel{100} \, \cancel{\text{mass AgCl}}} = 0.04065 \text{ g}$$

8-28. $\mathcal{M}_{AgCl} = 143.32 \text{ g/mol} \qquad \mathcal{M}_{AgI} = 234.77 \text{ g/mol}$

$$\text{mass AgCl} + \text{mass AgI} = 0.4430 \text{ g} \qquad (1)$$

$$\text{mass AgCl} + \text{mass AgI} \times \frac{1 \text{ mol AgI}}{234.77 \text{ g}} \times \frac{1 \text{ mol AgCl}}{1 \text{ mol AgI}} \times \frac{143.32 \text{ g}}{\text{mol AgCl}} = 0.3181 \text{ g}$$

or

$$\text{mass AgCl} + \text{mass AgI}\left(\frac{143.32}{234.77}\right) = 0.3181 \qquad (2)$$

By subtracting equation (2) from equation (1), we have

$$\text{mass AgI}\left(1 - \frac{143.32}{234.77}\right) = 0.1249 \text{ g}$$

$$\text{mass AgI} = \frac{0.1249 \text{ g}}{0.3895} = 0.3206 \text{ g}$$

Thus, mass AgCl = (0.4430 - 0.3206) = 0.1224 g

$$\frac{0.1224 \text{ g AgCl}}{0.6407 \text{ g sample}} \times \frac{1 \text{ mol AgCl}}{143.32 \text{ g}} \times \frac{1 \text{ mol Cl}^-}{\text{mol AgCl}} \times \frac{35.453 \text{ g}}{\text{mol Cl}^-} \times 100\% = \underline{\underline{4.72\% \text{ Cl}^-}}$$

$$\frac{0.3206 \text{ g AgI}}{0.6407 \text{ g sample}} \times \frac{1 \text{ mol AgI}}{234.77 \text{ g}} \times \frac{1 \text{ mol I}^-}{\text{mol AgI}} \times \frac{126.904 \text{ g}}{\text{mol I}^-} \times 100\% = \underline{\underline{27.05\% \text{ I}^-}}$$

	A	B	C	D
1	**Problem 8-28**			
2				
3	**Coefficient Matrix**			**Constant Matrix**
4	1	1		0.443
5	1	0.610469822		0.3181
6				
7	**Inverse Matrix**			**Solution Matrix**
8	-1.567195189	2.567195189		0.122357321
9	2.567195189	-2.56719519		0.320642679
10				
11	**Mass Sample**	**Mass AgCl**	**%Cl**	**%I**
12	0.6407	0.122357321	4.72	27.05
13				
14	**Spreadsheet Documentation**			
15	**A8:B9=MINVERSE(A4:B5)**			
16	**D8:D9=MMULT(A8:B9,D4:D5)**			
17	**B12=D4-D9**			
18	**C12=100*B12*35.453/(A12*143.32)**			
19	**D12=100*D9*126.904/(A12*234.77)**			

8-30. $\mathcal{M}_{CO_2} = 44.010$ g/mol

$\mathcal{M}_{MgCO_3} = 84.31$ g/mol

$\mathcal{M}_{K_2CO_3} = 138.21$ g/mol

$$\text{no. mol } CO_2 = \text{no. mol } MgCO_3 + \text{no. mol } K_2CO_3 =$$

$$1.204 \text{ g sample} \times \frac{36.0 \text{ g } MgCO_3}{100 \text{ g sample}} \times \frac{1 \text{ mol } MgCO_3}{84.31 \text{ g } MgCO_3} +$$

$$1.204 \text{ g sample} \times \frac{44.0 \text{ g } K_2CO_3}{100 \text{ g sample}} \times \frac{1 \text{ mol } K_2CO_3}{138.21 \text{ g } K_2CO_3}$$

$$\text{no. mol } CO_2 = (5.141 \times 10^{-3}) + (3.833 \times 10^{-3}) = 8.974 \times 10^{-3} \text{ mol}$$

$$\text{mass } CO_2 = 8.974 \times 10^{-3} \text{ mol} \times 44.010 \text{ g/mol} = \underline{\underline{0.395 \text{ g } CO_2}}$$

8-32. $\mathcal{M}_{BaCl_2 \cdot 2H_2O} = 244.26$ g/mol

$\mathcal{M}_{NaIO_3} = 197.89$ g/mol

$\mathcal{M}_{Ba(IO_3)_2} = 487.13$ g/mol

Initial amounts:

$$0.200 \text{ g } BaCl_2 \cdot H_2O \times \frac{\text{mol } BaCl_2 \cdot 2H_2O}{244.26 \text{ g}} \times \frac{1 \text{ mol } Ba^{2+}}{\text{mol } BaCl_2 \cdot 2H_2O} = 8.188 \times 10^{-4} \text{ mol } Ba^{2+}$$

$$0.300 \text{ g } NaIO_3 \times \frac{\text{mol } NaIO_3}{197.89 \text{ g}} \times \frac{1 \text{ mol } IO_3^-}{\text{mol } NaIO_3} = 1.5160 \times 10^{-3} \text{ mol } IO_3^-$$

Complete reaction of Ba^{2+} would require $(2 \times 8.188 \times 10^{-4}) = 1.6376 \times 10^{-3}$ mol of iodate ion. Since $1.5160 \times 10^{-3} < 1.6376 \times 10^{-3}$, barium ion is present in excess.

(a) $\text{no. mol Ba(IO}_3)_2 = \text{no. mol IO}_3^-/2 = 1.5160 \times 10^{-3}/2 = 7.580 \times 10^{-4}$

$$\text{mass Ba(IO}_3)_2 = 7.580 \times 10^{-4}\ \text{mol} \times \frac{487.13\ \text{g Ba(IO}_3)_2}{\text{mol}} = \underline{\underline{0.369\ \text{g Ba(IO}_3)_2}}$$

(b) $\text{no. mol BaCl}_2 \cdot 2\text{H}_2\text{O remaining} = (8.188 - 7.580) \times 10^{-4}\ \text{mol}$

$$= 6.08 \times 10^{-5}\ \text{mol}$$

$$\text{mass BaCl}_2 \cdot 2\text{H}_2\text{O} = 6.08 \times 10^{-5}\ \text{mol BaCl}_2 \cdot 2\text{H}_2\text{O} \times \frac{244.26\ \text{g BaCl}_2 \cdot 2\text{H}_2\text{O}}{\text{mol BaCl}_2 \cdot 2\text{H}_2\text{O}}$$

$$= \underline{\underline{0.0149\ \text{g BaCl}_2 \cdot 2\text{H}_2\text{O}}}$$

Chapter 9

9-1. **(a)** *Activity*, a, is the effective concentration of a species A in solution. The *activity coefficient*, γ_A, is the factor needed to convert a molar concentration to activity:

$$a_A = \gamma_A[A]$$

(b) The *thermodynamic equilibrium constant* refers to an ideal system within which each species is unaffected by any others. A *concentration equilibrium constant* takes account of the influence excerted by solute species upon one another. A thermodynamic constant is based upon activities of reactants and products; a concentration constant is based upon molar concentrations of reactants and products.

9-3. **(a)** Reaction:

$$MgCl_2 + 2NaOH \rightarrow Mg(OH)_2 + 2NaCl$$

The addition of NaOH has the effect of replacing a divalent ion (Mg^{2+}) with a chemically equivalent quantity of a univalent ion (Na^+); it should decrease the ionic strength.

(b) Reaction:

$$HCl + NaOH \rightarrow NaCl + H_2O$$

Addition of NaOH will convert HCl to an equivalent amount of NaCl. Because all of the ions involved are singly charged, μ should be unchanged.

(c) Reaction:

$$NaOH + HOAc \rightarrow NaOAc + H_2O$$

Addition of NaOH has the effect of replacing a slightly ionized species (HOAc) with a chemically equivalent quantity of water and Na^+; μ should increase.

9-5. For a given ionic strength, activity coefficients for ions with multiple charge show greater departures from ideality.

9-7. **(a)**

$$\mu = \frac{1}{2}[0.040(2)^2 + 0.040(2)^2] = \underline{\underline{0.16}}$$

(c) $\mu = \frac{1}{2}[0.10(3)^2 + 0.3(1)^2 + 0.20(2)^2 + 0.4(1)^2] = \underline{\underline{1.2}}$

9-8.

$$-\log \gamma_X = \frac{0.51\, Z_X^2 \sqrt{\mu}}{1 + 0.33\, \alpha_X \sqrt{\mu}}$$

	Z	μ	$\sqrt{\mu}$	α	$-\log \gamma$	$1/\gamma$	γ
(a)	3	0.075	0.274	9	0.693	4.93	0.20
(c)	4	0.080	0.283	11	1.139	13.77	0.073

9-9. (a) $\mu = 0.050 \quad \gamma_{Fe^{3+}} = 0.24$

$\mu = 0.10 \quad \gamma_{Fe^{2+}} = 0.18$

$$\gamma_{int} = 0.18 + \frac{0.025}{0.050}(0.24 - 0.18) = \underline{\underline{0.21}}$$

(c) $\mu = 0.05 \quad \gamma_{Ce^{4+}} = 0.10$

$\mu = 0.10 \quad \gamma_{Ce^{4+}} = 0.063$

$$\gamma_{int} = 0.063 + \frac{0.02}{0.05}(0.1 - 0.063) = \underline{\underline{0.078}}$$

9-10. (a) $\mu = 0.050 \quad \gamma_{Ag^+} = 0.80 \quad \gamma_{SCN^-} = 0.81$

$$K'_{sp} = \frac{K_{sp}}{\gamma_{Ag^+} \cdot \gamma_{SCN^-}} = \frac{1.1 \times 10^{-12}}{(0.80)(0.81)} = \underline{\underline{1.7 \times 10^{-12}}}$$

(c) $\gamma_{La^{3+}} = 0.24 \quad \gamma_{IO_3^-} = 0.82$

$$K'_{sp} = \frac{1.0 \times 10^{-11}}{0.24\,(0.82)^3} = \underline{\underline{7.6 \times 10^{-11}}}$$

9-11. $Zn(OH)_2(s) \rightleftarrows Zn^{2+} + 2OH^- \qquad K_{sp} = 3.0 \times 10^{-16}$

(a)

$$\mu = \frac{1}{2}[0.0100(1)^2 + 0.0100(1)^2] = 0.0100$$

In Table 9-1, we find at $\mu = 0.0100$

$$\gamma_{Zn^{2+}} = 0.676 \quad \text{and} \quad \gamma_{OH^-} = 0.900$$

$$K_{sp} = a_{Zn^{2+}} \cdot a^2_{OH^-} = 3.0 \times 10^{-16} = \gamma_{Zn^{2+}}[Zn^{2+}] \times \gamma^2_{OH^-}[OH^-]$$

$$[Zn^{2+}][OH^-]^2 = \frac{3.00 \times 10^{-16}}{\gamma_{Zn^{2+}} \cdot \gamma^2_{OH^-}} = \frac{3.00 \times 10^{-16}}{0.676(0.900)^2} = 5.48 \times 10^{-16} = K'_{sp}$$

$$\text{Solubility} = S = [Zn^{2+}] = \frac{1}{2}[OH^-]$$

$$S(2S)^2 = 5.48 \times 10^{-16}$$

$$S = \underline{\underline{5.2 \times 10^{-6}\ M}}$$

(b)

$$\mu = \frac{1}{2}[2 \times 0.0167(1)^2 + 0.0167(2)^2] = 0.050$$

$$\gamma_{Zn^{2+}} = 0.48 \qquad \gamma_{OH^-} = 0.81$$

$$K'_{sp} = \frac{3.0 \times 10^{-16}}{(0.48)(0.81)^2} = 9.53 \times 10^{-16}$$

$$S = [Zn^{2+}] = \left(\frac{9.53 \times 10^{-16}}{4}\right)^{1/3} = \underline{\underline{6.2 \times 10^{-6}\ M}}$$

(c) Initial amount

$$20.0 \times 0.250 = 5.00 \text{ mmol } K^+$$

$$= 5.00 \text{ mmol } OH^-$$

$$80.0 \times 0.0250 = 2.00 \text{ mmol } Zn^{2+}$$

$80.0 \times 0.0250 \times 2 \;=\; 4.00 \text{ mmol Cl}^-$

$c_{K^+} \;=\; 5/100 \;=\; 0.0500$

$c_{OH^-} \;=\; (5.00 - 2 \times 2.00)/100 \;=\; 0.0100$

$c_{Cl^-} \;=\; 0.0400$ and $c_{Zn^{2+}} \;=\; 0.00$

$$\mu \;=\; \frac{1}{2}[0.0500(1)^2 + 0.0100(1)^2 + 0.0400(1)^2] \;=\; 0.05$$

$\gamma_{Zn^{2+}} \;=\; 0.48$ and $\gamma_{OH^-} \;=\; 0.81$

$K'_{sp} \;=\; 9.53 \times 10^{-16}$ (see part b)

$[Zn^{2+}][OH^-]^2 \;=\; [Zn^{2+}](0.0100)^2 \;=\; 9.5 \times 10^{-16}$

$[Zn^{2+}] \;=\; S \;=\; \underline{\underline{9.5 \times 10^{-12}\text{ M}}}$

(d) Initial amount

$20.0 \times 0.100 \;=\; 2.00 \text{ mmol K}^+$

$\;=\; 2.00 \text{ mmol OH}^-$

$80.0 \times 0.0250 \;=\; 2.00 \text{ mmol Zn}^{2+}$

$80.0 \times 0.025033 \times 2 \;=\; 4.00 \text{ mmol Cl}^-$

$c_{K^+} \;=\; 2.00/100 \;=\; 0.0200\text{ M}$

$c_{OH^-} \;=\; 0.00$

$c_{Zn^{2+}} \;=\; (2.00 - 1.00)/100 \;=\; 0.0100$

$c_{Cl^-} \;=\; 4.00/100 \;=\; 0.0400$

$$\mu = \frac{1}{2}[0.0200\,(1)^2 + 0.0100\,(2)^2 + 0.0400\,(1)^2] = 0.05$$

$$K'_{sp} = 9.53 \times 10^{-16} = 0.0100\,[OH^-]^2 \qquad \text{(see part b)}$$

$$[OH^-] = \sqrt{9.53 \times 10^{-14}} = 3.09 \times 10^{-7}$$

$$S = [OH^-]/2 = \underline{\underline{1.5 \times 10^{-7}\text{ M}}}$$

9-12. $\mu = \frac{1}{2}\Sigma c_i Z_1^2 = \frac{1}{2}[0.0333(2)^2 + 0.0666(1)^2] = 0.100$ for all solutions

(a) (1) In Table 9-1 we find for $\mu = 0.100$,

$$\gamma_{Ag^+} = 0.75 \quad \text{and} \quad \gamma_{SCN^-} = 0.76$$

$$K_{sp} = a_{Ag^+} \times a_{SCN^-} = [Ag^+][SCN^-]\,\gamma_{Ag^+} \times \gamma_{SCN^-}$$

$$K'_{sp} = K_{sp}/(\gamma_{Ag^+} \times \gamma_{SCN^-}) = 1.1 \times 10^{-12}/(0.75 \times 0.76)$$

$$= 1.93 \times 10^{-12}$$

$$S = [Ag^+] = [SCN^-]$$

$$S = \sqrt{K'_{sp}} = \sqrt{1.93 \times 10^{-12}} = \underline{\underline{1.4 \times 10^{-6}\text{ M}}}$$

(2) $S = \sqrt{K_{sp}} = \sqrt{1.1 \times 10^{-12}} = \underline{\underline{1.0 \times 10^{-6}\text{ M}}}$

(b) (1) $\gamma_{Pb^{2+}} = 0.36$ and $\gamma_{I^-} = 0.75$

$$K_{sp} = [Pb^{2+}][I^-]\,\gamma_{Pb^{2+}} \cdot \gamma_{I^-}^2$$

$$K'_{sp} = K_{sp}/\left(\gamma_{Pb^{2+}} \cdot \gamma_{I^-}^2\right) = 7.9 \times 10^{-9}/[0.36 \times (0.75)^2] = 3.90 \times 10^{-8}$$

$$[Pb^{2+}] = S \quad \text{and} \quad [I^-] = 2\,S$$

$$S(2S)^2 = 3.90 \times 10^{-8}$$

$$S = (3.90 \times 10^{-8}/4)^{1/3} = \underline{\underline{2.1 \times 10^{-3}\ \text{M}}}$$

(2) $$S = (7.9 \times 10^{-9}/4)^{1/3} = \underline{\underline{1.3 \times 10^{-3}\ \text{M}}}$$

(c) (1) Proceeding as in part (a),

$$K'_{sp} = 1.1 \times 10^{-10}/(0.38 \times 0.35) = 8.27 \times 10^{-10}$$

$$S = [Ba^{2+}] = [SO_4^{2-}]$$

$$S = \sqrt{K'_{sp}} = \sqrt{8.27 \times 10^{-10}} = \underline{\underline{2.9 \times 10^{-5}\ \text{M}}}$$

(2) $$S = \sqrt{1.1 \times 10^{-10}} = \underline{\underline{1.0 \times 10^{-5}\ \text{M}}}$$

(d) (1) Proceeding as in part (b),

$$K'_{sp} = 3.2 \times 10^{-17}/[(0.38)^2 \times 0.020] = 1.11 \times 10^{-14}$$

$$[Cd^{2+}] = 2S \quad \text{and} \quad [Fe(CN)_6^{4-}] = S$$

$$S(2S)^2 = 4S^3 = 1.06 \times 10^{-14}$$

$$S = (1.11 \times 10^{-14}/4)^{1/3} = \underline{\underline{1.4 \times 10^{-5}\ \text{M}}}$$

(2) $$S = (3.2 \times 10^{-17}/4)^{1/3} = \underline{\underline{2.0 \times 10^{-6}\ \text{M}}}$$

9-13. $$\mu = \frac{1}{2}[0.0167(2)^2 + 2 \times 0.0167(1)^2] = 0.05$$

(a) (1) $$\gamma_{Ag^+} = 0.80 \quad \text{and} \quad \gamma_{IO_3^-} = 0.77$$

$$K'_{sp} = \frac{3.1 \times 10^{-8}}{0.80 \times 0.77} = 5.03 \times 10^{-8}$$

$$S = \sqrt{5.03\times10^{-8}} = \underline{\underline{2.2\times10^{-4}\ \text{M}}}$$

(2) $$S = \sqrt{3.1\times10^{-8}} = \underline{\underline{1.8\times10^{-4}\ \text{M}}}$$

(b) (1) $\gamma_{Mg^{2+}} = 0.52$ and $\gamma_{OH^-} = 0.81$

$$K'_{sp} = \frac{7.1\times10^{-12}}{0.52\,(0.81)^2} = 2.08\times10^{-11}$$

$$S = [Mg^{2+}] = \frac{1}{2}[OH^-]$$

$$S(2S)^2 = 2.08\times10^{-11}$$

$$S = (2.08\times10^{-11}/4)^{1/3} = \underline{\underline{1.7\times10^{-4}\ \text{M}}}$$

(2) $$S = (7.1\times10^{-12}/4)^{1/3} = \underline{\underline{1.2\times10^{-4}\ \text{M}}}$$

(c) (1) $\gamma_{Ba^{2+}} = 0.46$ and $\gamma_{SO_4^{2-}} = 0.44$

$$K'_{sp} = \frac{1.1\times10^{-10}}{0.46\times0.44} = 5.44\times10^{-10}$$

$$S = 5.43\times10^{-10}/0.0167 = \underline{\underline{3.3\times10^{-8}\ \text{M}}}$$

(2) $$S = 1.1\times10^{-10}/0.0167 = \underline{\underline{6.6\times10^{-9}\ \text{M}}}$$

(d) (1) $\gamma_{La^{3+}} = 0.24$ and $\gamma_{IO_3^-} = 0.82$

$$K'_{sp} = \frac{1.0\times10^{-11}}{0.24\,(0.82)^3} = 7.56\times10^{-11}$$

$$S = [La^{3+}] = \frac{1}{3}[IO_3^-]$$

$$S(3S)^3 = 7.56 \times 10^{-11}$$

$$S = (7.56 \times 10^{-11}/27)^{1/4} = \underline{\underline{1.3 \times 10^{-3}\ \text{M}}}$$

(2) $$S = (1.0 \times 10^{-11}/27)^{1/4} = \underline{\underline{8 \times 10^{-4}\ \text{M}}}$$

Chapter 10

10-3. A charge-balance equation is derived by relating the concentration of cations and anions in such a way that

no. mol/L positive charge = no. mol/L negative charge

For a doubly charged ion such as Ba^{2+}, the concentration of electrons for each mole is twice the molar *concentration* of the Ba^{2+}. That is,

$$\text{mol/L positive charge} = 2[Ba^{2+}]$$

Thus the molar concentration of all multiply charged species is always multiplied by the charge in a charge-balance equation.

10-4. **(a)** $0.10 = [H_3PO_4] + [H_2PO_4^-] + [HPO_4^{2-}] + [PO_4^{3-}]$

(c) $0.100 + 0.0500 = [HNO_2] + [NO_2^-]$

$$[Na^+] = c_{NaNO_2} = 0.0500$$

(e) $0.100 = [Na^+] = [OH^-] + 2[Zn(OH)_4^{2-}]$

(g) $[Ca^{2+}] = \frac{1}{2}([F^-] + [HF])$

10-5. **(a)** $[H_3O^+] = [OH^-] + [H_2PO_4^-] + 2[HPO_4^{2-}] + 3[PO_4^{3-}]$

(c) $[Na^+] + [H_3O^+] = [OH^-] + [NO_2^-]$

(e) $2[Zn^{2+}] + [H_3O^+] + [Na^+] = 2[Zn(OH)_4^{2-}] + [OH^-]$

(g) $2[Ca^{2+}] + [H_3O^+] = [OH^-] + [F^-]$

10-6. **(a)** Following the systematic procedure shown in Figure 10-1 and Example 10-6, we write

Step 1 $Ag_2CO_3(s) \rightleftarrows 2Ag^+ + CO_3^{2-}$

$$H_2CO_3 + H_2O \rightleftarrows H_3O^+ + HCO_3^-$$

$$HCO_3^- + H_2O \rightleftarrows H_3O^+ + CO_3^{2-}$$

Step 2 $S \;=\; \text{solubility} \;=\; [Ag^+]/2$

Step 3 $[Ag^+]^2[CO_3^{2-}] \;=\; K_{sp} \;=\; 8.1\times10^{-12}$ (1)

$$\frac{[H_3O^+][HCO_3^-]}{[H_2CO_3]} \;=\; K_1 \;=\; 1.5\times10^{-4} \qquad (2)$$

$$\frac{[H_3O^+][CO_3^{2-}]}{[HCO_3^-]} \;=\; K_2 \;=\; 4.69\times10^{-11} \qquad (3)$$

Step 4 $[Ag^+] \;=\; 2([CO_3^{2-}] + [HCO_3^-] + [H_2CO_3])$ (4)

$$[H_3O^+] \;=\; 1.00\times10^{-6}$$

Step 5 No charge-balance equation because a buffer of unknown composition is present.

Step 6 Unknowns: $[Ag^+]$, $[CO_3^{2-}]$, $[HCO_3^-]$, and $[H_2CO_3]$

Equations: (1), (2), (3), and (4)

Step 7 No approximations needed.

Step 8 Substituting $[H_3O^+] = 1.00\times10^{-6}$ into equation (3) and rearranging gives

$$[HCO_3^-] \;=\; \frac{1.00\times10^{-6}[CO_3^{2-}]}{4.69\times10^{-11}} \;=\; 2.13\times10^{4}[CO_3^{2-}]$$

Substituting this relationship and $[H_3O^+]$ into equation (2) gives after rearranging

$$[H_2CO_3] \;=\; \frac{1.00\times10^{-6}\times2.13\times10^{4}[CO_3^{2-}]}{1.5\times10^{-4}} \;=\; 1.42\times10^{2}\,[CO_3^-]$$

Substituting these two relationships into equation (4) gives

$$[Ag^+] = 2[CO_3^{2-}] + 2\times 2.13\times 10^4[CO_3^{2-}] + 2\times 1.42\times 10^2[CO_3^{2-}]$$

$$= 4.29\times 10^4[CO_3^{2-}]$$

Substituting into equation (1) gives

$$[Ag^+] = \frac{4.29\times 10^4\times 8.1\times 10^{-12}}{[Ag^+]^2}$$

or $[Ag^+] = (4.29\times 10^4\times 8.1\times 10^{-12})^{1/3} = 7.0\times 10^{-3}$

$$\text{solubility} = 7.0\times 10^{-3}/2 = \underline{\underline{3.5\times 10^{-3}\text{ M}}}$$

Substituting other values for $[H_3O^+]$ yields the following solubility data:

	$[H_3O^+]$	Solubility, M
(a)	1.00×10^{-6}	3.5×10^{-3}
(c)	1.00×10^{-9}	3.6×10^{-4}

10-7. Proceeding as in Solution 10-6, we write

$$BaSO_4^{2-} \rightleftarrows Ba^{2+} + SO_4^{2-} \qquad K_{sp} = 1.1\times 10^{-10}$$

$$HSO_4^- + H_2O \rightleftarrows H_3O^+ + SO_4^{2-} \qquad K_2 = 1.02\times 10^{-2}$$

$$S = [Ba^{2+}]$$

$$[Ba^{2+}][SO_4^{2-}] = 1.1\times 10^{-10} \qquad (1)$$

$$\frac{[H_3O^+][SO_4^{2-}]}{[HSO_4^-]} = 1.02\times 10^{-2} \qquad (2)$$

Mass balance requires that

$$[Ba^{2+}] = [SO_4^{2-}] + [HSO_4^-] \qquad (3)$$

Since $[H_3O^+]$ is known, we have 3 equations and 3 unknowns.

Substituting equation (2) into equation (3) yields

$$[Ba^{2+}] = [SO_4^{2-}] + \frac{[H_3O^+][SO_4^{2-}]}{1.02 \times 10^{-2}}$$

$$= [SO_4^{2-}](1 + [H_3O^+]/1.02 \times 10^{-2})$$

Substituting equation (1) to eliminate $[SO_4^-]$ give

$$[Ba^{2+}] = \frac{1.1 \times 10^{-10}}{[Ba^{2+}]} \times (1 + 98.0[H_3O^+])$$

$$[Ba^{2+}] = S = \sqrt{1.1 \times 10^{-10}(1 + 98.0[H_3O^+])}$$

$$= \sqrt{1.1 \times 10^{-10} + 1.078 \times 10^{-8}[H_3O^+]}$$

Substituting for $[H_3O^+]$ in this equation leads to

	$[H_3O^+]$	S, mol/L
(a)	2.00	1.47×10^{-4}
(c)	0.500	7.42×10^{-5}

10-8. The following derivation applies to this and the following two problems.

$$MS(s) \rightleftarrows M^{2+} + S^{2-}$$

$$H_2S + H_2O \rightleftarrows H_3O^+ + HS^- \qquad K_1 = 9.6 \times 10^{-8}$$

$$HS^- + H_2O \rightleftarrows H_3O^+ + S^{2-} \qquad K_2 = 1.3 \times 10^{-14}$$

$$H_2S + 2H_2O \rightleftarrows 2H_3O^+ + S^{2-} \qquad K_1K_2 = 9.6 \times 10^{-8} \times 1.3 \times 10^{-14} = 1.25 \times 10^{21}$$

$$S = \text{solubility} = [M^{2+}] = [S^{2-}] + [HS^-] + [H_2S]$$

$$[M^{2+}][S^{2-}] = K_{sp} \qquad (1)$$

$$\frac{[H_3O^+][S^{2-}]}{[HS^-]} = K_2 = 1.3\times10^{-14} \qquad (2)$$

$$\frac{[H_3O^+]^2[S^{2-}]}{[H_2S]} = K_1K_2 = 1.25\times10^{-21} \qquad (3)$$

From mass-balance consideration

$$[M^{2+}] = [S^{2-}]+[HS^-]+[H_2S] \qquad (4)$$

Substituting equations (2) and (3) into (4) gives

$$[M^{2+}] = [S^{2-}]+\frac{[H_3O^+][S^{2-}]}{K_2}+\frac{[H_3O^+]^2[S^{2-}]}{K_1K_2}$$

$$= [S^{2-}]\left(1+\frac{[H_3O^+]}{K_2}+\frac{[H_3O^+]^2}{K_1K_2}\right)$$

Substituting equation (1) yields

$$[M^{2+}] = \frac{K_{sp}}{[M^{2+}]}\left(1+\frac{[H_3O^+]}{K_2}+\frac{[H_3O^+]^2}{K_1K_2}\right)$$

$$[M^{2+}] = \sqrt{K_{sp}\left(1+\frac{[H_3O^+]}{K_2}+\frac{[H_3O^+]^2}{K_1K_2}\right)} \qquad (5)$$

$$= \sqrt{K_{sp}\left(1+\frac{[H_3O^+]}{1.3\times10^{-14}}+\frac{[H_3O^+]^2}{1.25\times10^{-21}}\right)}$$

(a) Substituting $K_{sp} = 8\times10^{-37}$ and $[H_3O^+] = 0.10$ into equation (5) leads to

$$[Cu^{2+}] = \text{solubility} = \sqrt{8\times10^{-37}\left(1+\frac{0.10}{1.3\times10^{-14}}+\frac{(0.10)^2}{1.25\times10^{-21}}\right)}$$

$$= \underline{\underline{2.5\times10^{-9}\,M}}$$

(b) Substituting $K_{sp} = 8\times10^{-37}$ and $[H_3O^+] = 1.0\times10^{-4}$ into equation (5) gives

$$\text{solubility} = \underline{\underline{2.5 \times 10^{-12}\ \text{M}}}$$

10-11. Here we proceed as in Solution 10-6.

$$PbCO_3 \rightleftarrows Pb^{2+} + CO_3^{2-} \qquad K_{sp} = [Pb^{2+}][CO_3^{2-}] = 7.4 \times 10^{-14} \qquad (1)$$

$$H_2CO_3 + H_2O \rightleftarrows H_3O^+ + HCO_3^- \qquad K_1 = \frac{[H_3O^+][HCO_3^-]}{[H_2CO_3]} = 1.5 \times 10^{-4} \qquad (2)$$

$$HCO_3^- + H_2O \rightleftarrows H_3O^+ + CO_3^{2-} \qquad K_2 = \frac{[H_3O^+][CO_3^{2-}]}{[HCO_3^-]} = 4.69 \times 10^{-11} \qquad (3)$$

$$[Pb^{2+}] = [CO_3^{2-}] + [HCO_3^-] + [H_2CO_3]$$

$$[H_3O^+] = 1.00 \times 10^{-7}$$

Proceeding as in Solution 10-8 we derive an equation analogous to equation (5) in that solution. That is,

$$[Pb^{2+}] = \sqrt{K_{sp}\left(1 + \frac{[H_3O^+]}{K_2} + \frac{[H_3O^+]^2}{K_1K_2}\right)}$$

$$= \sqrt{7.4 \times 10^{-14}\left(1 + \frac{1.0 \times 10^{-7}}{4.69 \times 10^{-11}} + \frac{1.0 \times 10^{-14}}{1.5 \times 10^{-4} \times 4.69 \times 10^{-11}}\right)}$$

$$\text{solubility} = [Pb^{2+}] = \underline{\underline{1.3 \times 10^{-5}\ \text{M}}}$$

10-13. $[Cu^{2+}][OH^-]^2 = 4.8 \times 10^{-20}$ $\qquad [Mn^{2+}][OH^-]^2 = 2 \times 10^{-13}$

(a) $\underline{\underline{\text{Cu(OH)}_2 \text{ forms first}}}$

(b) Cu^{2+} begins to precipitate when

$$[OH^-] = \sqrt{4.8 \times 10^{-20}/0.050} = \underline{\underline{9.8 \times 10^{-10}}}$$

(c) Mn^{2+} begins to precipitate when

$$[OH^-] = \sqrt{2 \times 10^{-13}/0.0400} = 2.2 \times 10^{-6}$$

$$[Cu^{2+}] = 4.8 \times 10^{-20}/(2.2 \times 10^{-6})^2 = \underline{\underline{9.6 \times 10^{-9}}}$$

10-15. (a) $[Ag^+] = 8.3 \times 10^{-17}/1.0 \times 10^{-6} = \underline{\underline{8.3 \times 10^{-11}}}$

(b) $[Ag^+] = 1.1 \times 10^{-12}/0.070 = \underline{\underline{1.6 \times 10^{-11}}}$

(c) $[I^-] = 8.3 \times 10^{-17}/1.6 \times 10^{-11} = 5.2 \times 10^{-6}$

$$[SCN^-]/[I^-] = 0.070/5.2 \times 10^{-6} = \underline{\underline{1.3 \times 10^4}}$$

(d) $\dfrac{[SCN^-]}{[I^-]} = \dfrac{1.1 \times 10^{-12}/[Ag^+]}{8.3 \times 10^{-17}/[Ag^+]} = \underline{\underline{1.3 \times 10^4}}$

Note that this ratio is independent of $[Ag^+]$ as long as some AgSCN(*s*) and AgI(*s*) is present.

10-17. $AgBr(s) \rightleftarrows Ag^+ + Br^- \qquad K_{sp} = 5.0 \times 10^{-13} = [Ag^+][Br^-] \qquad (1)$

$$Ag^+ + 2CN^- \rightleftarrows Ag(CN)_2^- \qquad \beta_2 = \frac{[Ag(CN)_2^-]}{[Ag^+][CN^-]^2} = 1.3 \times 10^{21} \qquad (2)$$

It is readily shown that the reaction

$$CN^- + H_2O \rightleftarrows HCN + OH^-$$

proceeds to such a small extent that it can be neglected in formulating a solution to this problem. That is, $HCN << [CN^-]$, and only the two equilibria shown need to be taken into account.

$$\text{Solubility} = [Br^-] = [Ag^+] + [Ag(CN)_2^-]$$

Mass balance requires that

$$[Br^-] = [Ag^+] + [Ag(CN)_2^-] \qquad (3)$$

$$0.100 = [CN^-] + 2[Ag(CN)_2^-] \qquad (4)$$

We now have 4 equations and 4 unknowns.

Based upon the large size of β_2, let us assume that

$$[CN^-] << [2\,Ag(CN)_2^-] \qquad \text{and} \qquad [Ag^+] << [Ag(CN)_2^-]$$

Equation (4) becomes

$$[Ag(CN)_2^-] = 0.100/2 = 0.0500$$

and equation (3) becomes

$$[Br^-] = [Ag(CN)_2^-] = 0.0500$$

To check the assumption, we calculate $[Ag^+]$ by substituting into equation (1)

$$[Ag^+] = 5.0 \times 10^{-13}/0.0500 = 1 \times 10^{-11}$$

To obtain $[CN^-]$ we substitute into equation (2)

$$\frac{[Ag(CN)_2^-]}{[Ag^+][CN^-]^2} = \frac{0.0500}{1 \times 10^{-11}[CN^-]^2} = 1.3 \times 10^{21}$$

$$[CN^-] = 2.0 \times 10^{-6}$$

Thus, the two assumptions are valid and

$$\text{solubility} = [Br^-] = 0.0500 \text{ M}$$

$$\text{mass AgBr}/200 \text{ mL} = 0.0500 \frac{\text{mmol}}{\text{mL}} \times 200 \text{ mL} \times \frac{0.1877 \text{ g AgBr}}{\text{mmol AgBr}} = \underline{\underline{1.877 \text{ g}}}$$

10-19. (a) $CaSO_4(s) \rightleftarrows Ca^{2+} + SO_4^{2-}$ $\quad K_{sp} = 2.4 \times 10^{-5} = [Ca^{2+}][SO_4^{2-}]$ $\quad$ (1)

$CaSO_4(aq) \rightleftarrows Ca^{2+} + SO_4^{2-}$ $\quad K_d = 5.2 \times 10^{-3} = \dfrac{[Ca^{2+}][SO_4^{2-}]}{[CaSO_4]_{aq}}$ $\quad$ (2)

$CaSO_4(s) \rightleftarrows CaSO_4(aq)$

$[Ca^{2+}] = [SO_4^{2-}]$ $\quad$ (3)

Three equations and 3 unknowns ($[Ca^{2+}]$, $[SO_4^{2-}]$, and $[CaSO_4]_{aq}$).

To solve the equation we divide equation (1) by (2) to give

$$[CaSO_4]_{aq} = K_{sp}/K_d = 2.4 \times 10^{-5}/(5.2 \times 10^{-3}) = 4.6 \times 10^{-3}$$

Note that this is the equilibrium constant expression for the third equilibrium and indicates the concentration of unionized $CaSO_4$ *is always the same in a saturated solution of CaSO₄*. Substituting equation (3) into (1) gives

$$[Ca^{2+}] = \sqrt{2.4 \times 10^{-5}} = 4.90 \times 10^{-3}$$

$$S = [CaSO_4]_{aq} + [Ca^{2+}]$$

$$= 4.6 \times 10^{-3} + 4.9 \times 10^{-3} = 9.5 \times 10^{-3} = \underline{\underline{0.0095 \text{ M}}}$$

$$\% \ CaSO_4(aq) = (4.6 \times 10^{-3}/9.5 \times 10^{-3}) \times 100\% = \underline{\underline{48\%}}$$

(b) Here, $[CaSO_4]_{aq}$ is again equal to 4.6×10^{-3}

$$[SO_4^{2-}] = 0.0100 + [Ca^{2+}] \approx 0.0100$$

$$[Ca^{2+}] = 2.4 \times 10^{-5}/0.0100 = 2.4 \times 10^{-3}$$

Thus, the approximation is not very good and we write

$$[Ca^{2+}] = K_{sp}/[SO_4^{2-}] = 2.4 \times 10^{-5}/(0.0100 + [Ca^{2+}])$$

We solve this equation by systematic approximations. Thus, if $[Ca^{2+}] = 0.00$

$$[Ca^{2+}]_1 = 2.4 \times 10^{-5} / (0.0100 + 0.00) = 2.4 \times 10^{-3}$$

$$[Ca^{2+}]_2 = 2.4 \times 10^{-5} / (0.0100 + 2.4 \times 10^{-3}) = 1.94 \times 10^{-3}$$

$$[Ca^{2+}]_3 = 2.4 \times 10^{-5} / (0.0100 + 1.94 \times 10^{-3}) = 2.01 \times 10^{-3}$$

$$[Ca^{2+}] = 2.4 \times 10^{-5} / (0.0100 + 2.01 \times 10^{-3}) = 2.00 \times 10^{-3}$$

$$S = 4.6 \times 10^{-3} + 2.00 \times 10^{-3} = \underline{\underline{6.6 \times 10^{-3}\ \text{M}}}$$

$$\%\ \text{CaSO}_4(aq) = (4.6 \times 10^{-3} / (6.6 \times 10^{-3}) \times 100\% = \underline{\underline{70\%}}$$

Chapter 11

11-2. **(a)** The *millimole* is the amount of an elementary species, such as an atom, an ion, a molecule, or an electron. A millimole contains

$$6.02\times10^{23}\,\frac{\text{particles}}{\cancel{\text{mole}}}\times10^{-3}\,\frac{\cancel{\text{mole}}}{\text{millimole}} = 6.02\times10^{20}\,\frac{\text{particles}}{\text{millimole}}$$

(c) The *stoichiometric ratio* is the molar ratio of two species that appear in a balanced chemical equation.

11-3. **(a)** The *equivalence point* in a titration is that point at which sufficient titrant has been added so that stoichiometrically equivalent amounts of analyte and titrant are present. The *end point* in a titration is the point at which an observable physical change signals the equivalence point.

(c) A *primary standard* is a highly purified substance that serves as the basis for a titrimetric method. It is used either (1) to prepare a standard solution directly by mass or (2) to standardize a solution to be used in a titration.

A *secondary standard* is material or solution whose concentration is determined from the stoichiometry of its reaction with a primary standard material. Secondary standards are employed when a reagent is not available in primary standard quality. For example, solid sodium hydroxide is hygroscopic and cannot be used to prepare a standard solution directly. A secondary standard solution of the reagent is readily prepared, however, by standardizing a solution of sodium hydroxide against a primary standard reagent such as potassium hydrogen phthalate.

11-4. For a dilute aqueous solution $1\text{ L} = 1000\text{ mL} = 1000\text{ g}$, so

$$\frac{\text{mg}}{\text{L}} = \frac{10^{-3}\text{ g solute}}{1000\text{ g soln}} = \frac{1\text{ g solute}}{1{,}000{,}000\text{ g soln}} = 1\text{ ppm}$$

11-5. **(a)** $\dfrac{1\text{ mol}H_2NNH_2}{2\text{ mol }I_2}$ **(c)** $\dfrac{1\text{ mol }Na_2B_4O_7\cdot10H_2O}{2\text{ mol }H^+}$

11-6. **(a)** $$2.00\text{ L}\times2.76\times10^{-3}\,\frac{\text{mol}}{\text{L}}\times\frac{1000\text{ mmol}}{\text{mol}} = \underline{\underline{5.52\text{ mmol}}}$$

(b) $750\text{ mL} \times \dfrac{0.0416\text{ mmol}}{\text{mL}} = \underline{\underline{31.2\text{ mmol}}}$

(c) $\dfrac{4.20\text{ g CuSO}_4}{10^6\text{ g soln}} \times \dfrac{1.00\text{ g soln}}{\text{mL soln}} \times \dfrac{1\text{ mmol CuSO}_4}{0.1596/\text{g CuSO}_4} \times 250\text{ mL soln} = \underline{\underline{6.58 \times 10^{-3}\text{ mmol}}}$

(d) $3.50\text{ L} \times 0.276 \dfrac{\text{mol}}{\text{L}} \times \dfrac{1000\text{ mmol}}{\text{mol}} = \underline{\underline{966\text{ mmol}}}$

11-8. **(a)** $26.0\text{ mL} \times \dfrac{0.150\text{ mmol sucrose}}{\text{mL}} \times \dfrac{0.342\text{ g}}{\text{mmol sucrose}} \times \dfrac{1000\text{ mg}}{\text{g}} = \underline{\underline{1.33 \times 10^3\text{ mg}}}$

(b) $2.92\text{ L} \times \dfrac{5.23 \times 10^{-3}\text{ mol H}_2\text{O}_2}{\text{L}} \times \dfrac{34.02\text{ g}}{\text{mol H}_2\text{O}_2} \times \dfrac{1000\text{ mg}}{\text{g}} = \underline{\underline{520\text{ mg}}}$

11-9. **(a)** $450\text{ mL} \times \dfrac{0.164\text{ mol H}_2\text{O}_2}{\text{L}} \times \dfrac{34.02\text{ g}}{\text{molH}_2\text{O}_2} \times \dfrac{1\text{ L}}{1000\text{ mL}} = \underline{\underline{2.51\text{ g}}}$

(b) $27.0\text{ mL} \times \dfrac{8.75 \times 10^{-5}\text{ mol}}{\text{L}} \times \dfrac{122.1\text{ g}}{\text{mol}} \times \dfrac{1\text{ L}}{1000\text{ mL}} = \underline{\underline{2.88 \times 10^{-3}\text{ g}}}$

11-10. $\dfrac{50.0\text{ g NaOH}}{100\text{ g soln}} \times \dfrac{1.52\text{ g soln}}{\text{mL soln}} \times \dfrac{1\text{ mmol NaOH}}{0.0400\text{ g NaOH}} = 19.0 \dfrac{\text{mmol NaOH}}{\text{mL soln}} = 19.0\text{ M}$

11-12. Proceeding as in Solution 11-10 we obtain

(a) 6.161 M **(c)** 4.669 M

11-13. **(a)** $500\text{ mL soln} \times \dfrac{16.0\text{ g EtOH}}{100\text{ mL soln}} = 80.0\text{ g EtOH}$

Dilute 80 g ethanol to 500 mL with water.

(b) $500\text{ mL soln} \times \dfrac{16.0\text{ mL EtOH}}{100\text{ mL soln}} = 80.0\text{ mL EtOH}$

Dilute 80.0 mL ethanol to 500 mL with water.

(c) $500 \text{ g soln} \times \frac{16.0 \text{ g EtOH}}{100 \text{ g soln}} = 80.0 \text{ g EtOH}$

Dilute 80.0 g ethanol with 420 g water.

11-15. $c_{HClO_4} = \frac{1.66 \times 10^3 \text{g reagent}}{\text{L reagent}} \times \frac{70 \text{ g HClO}_4}{100 \text{ g reagent}} \times \frac{1 \text{ mol HClO}_4}{100.5 \text{ g HClO}_4} = 11.56 \frac{\text{mol HClO}_4}{\text{L reagent}}$

$$\text{Amount HClO}_4 \text{ required} = 2.00 \text{ L} \times \frac{0.150 \text{ mol HClO}_4}{\text{L}} = 0.300 \text{ mol HClO}_4$$

$$\text{Vol concd reagent} = 0.300 \text{ mol HClO}_4 \times \frac{1 \text{ L reagent}}{11.56 \text{ mol HClO}_4}$$

$$= 0.0260 \text{ L} \quad \text{or} \quad 26.0 \text{ mL}$$

Dilute 26 mL of the concentrated reagent to 2.0 L.

11-17. (a) $500 \text{ mL} \times \frac{1 \text{ L}}{1000 \text{ mL}} \times \frac{0.0750 \text{ mol}}{1 \text{ L}} \times \frac{169.87 \text{ g}}{1 \text{ mol}} = 6.37 \text{ g}$

Dissolve 6.37 g of $AgNO_3$ in water and dilute to 500 mL.

(b) $V_{concd} c_{concd} = V_{dil} c_{dil}$

$$V_{concd} = \frac{V_{dil} c_{dil}}{c_{concd}} = \frac{1.00 \text{ L} \times 0.315 \text{ M}}{6.00 \text{ M}} = 0.0525 \text{ L} = 52.5 \text{ mL}$$

Dilute 52.5 mL of 6.00 M HCl to 1.00 L.

(c) $600 \text{ mL} \times \frac{1 \text{ L}}{1000 \text{ mL}} \times \frac{0.0825 \text{ mol K}^+}{\text{L}} \times \frac{1 \text{ mol K}_4\text{Fe(CN)}_6}{4 \text{ mol K}^+} \times \frac{368.35 \text{ g}}{1 \text{ mol K}_2\text{Fe(CN)}_6} = 4.56 \text{ g}$

Dissolve 4.56 g $K_4Fe(CN)_6$ in water and dilute to 600 mL.

(d) $400 \text{ mL} \times \frac{3.00 \text{ g}}{100 \text{ mL}} \times \frac{1 \text{ mol}}{208.23 \text{ g}} \times \frac{1 \text{ L}}{0.400 \text{ mol}} = 0.144 \text{ L} = 144 \text{ mL}$

Dilute 144 mL of 0.400 M $BaCl_2$ to a volume of 400 mL.

(e) $$c_{HClO_4} = \frac{1.60\times10^3 \text{ reagent}}{\text{L reagent}}\times\frac{60 \text{ g HClO}_4}{100 \text{ g reagent}}\times\frac{1 \text{ mol HClO}_4}{100.5 \text{ g HClO}_4} = 9.55\frac{\text{mol HClO}_4}{\text{L reagent}}$$

$$\text{Amount HClO}_4 \text{ reguired} = 2.00 \text{ L}\times0.120\frac{\text{mol HClO}_4}{\text{L}} = 0.240 \text{ mol HClO}_4$$

$$\text{Vol concd reagent} = 0.240 \text{ mol HClO}_4\times\frac{1 \text{ mol reagent}}{9.55 \text{ mol HClO}_4}\times\frac{10^3\text{mL}}{\text{L}} = 25.1 \text{ mL}$$

Dilute 25 mL of the commercial reagent to a volume of 2.0 L.

(f) $$9.00\times10^3 \text{ mL soln}\times\frac{1.00 \text{ g soln}}{\text{mL soln}}\times\frac{60 \text{ g Na}^+}{10^6 \text{ g soln}}\times\frac{1 \text{ mol Na}^+}{22.99 \text{ g soln}}\times\frac{1 \text{ mol Na}_2\text{SO}_4}{2 \text{ mol Na}^+}\times$$

$$\frac{142.0 \text{ g Na}_2\text{SO}_4}{\text{mol Na}_2\text{SO}_4} = 1.67 \text{ g Na}_2\text{SO}_4$$

Dissolve 1.67 g Na_2SO_4 in water and dilute to 9.00 L.

11-19. $\mathcal{M}_{HgO} = 216.59 \text{ g/mol}$

$$\frac{0.3745 \text{ g HgO}\times\frac{1 \text{ mmol HgO}}{0.21659 \text{ g HgO}}\times\frac{2 \text{ mmol OH}^-}{1 \text{ mmol HgO}}\times\frac{1 \text{ mmol HClO}_4}{1 \text{ mmol OH}^-}}{37.79 \text{ mL soln}} = 9.151\times10^{-2} \text{ M HClO}_4$$

11-21. $\mathcal{M}_{Na_2SO_4} = 142.04 \text{ g/mol}$

$$\frac{0.3396 \text{ g sample}\times\frac{96.4 \text{ g Na}_2\text{SO}_4}{100 \text{ g sample}}\times\frac{1 \text{ mmol Na}_2\text{SO}_4}{0.14204 \text{ g Na}_2\text{SO}_4}\times\frac{1 \text{ mmol BaCl}_2}{1 \text{ mmol Na}_2\text{SO}_4}}{37.70 \text{ mL}} = 0.06114 \text{ M BaCl}_2$$

11-22. $$\frac{V_{HClO_4}}{V_{NaOH}} = \frac{27.43 \text{ mL HClO}_4}{25.00 \text{ mL NaOH}} = 1.0972\frac{\text{mL HClO}_4}{\text{mL NaOH}}$$

The volume of $HClO_4$ needed to titrate 0.4793 g of Na_2CO_3 is

$$40.00 \text{ mL HClO}_4 - 8.70 \text{ mL NaOH}\times\frac{1.0972 \text{ mL HClO}_4}{\text{mL NaOH}} = 30.45 \text{ mL}$$

Thus

$$\frac{0.4793\text{ g Na}_2\text{CO}_3}{30.45\text{ mL HClO}_4}\times\frac{1\text{ mmol Na}_2\text{CO}_3}{0.10599\text{ g Na}_2\text{CO}_3}\times\frac{2\text{ mmol HClO}_4}{\text{mmol Na}_2\text{CO}_3} = \underline{\underline{0.2970\text{ M HClO}_4}}$$

and

$$c_{\text{NaOH}} = c_{\text{HClO}_4}\times\frac{V_{\text{HClO}_4}}{V_{\text{NaOH}}}$$

$$= \frac{0.2970\text{ mmol HClO}_4}{\text{mL HClO}_4}\times\frac{1.0972\text{ mL HClO}_4}{\text{mL NaOH}}\times\frac{1\text{ mmol NaOH}}{\text{mmol HClO}_4} = \underline{\underline{0.3259\text{ M}}}$$

11-24.

$$\frac{0.1238\text{ g KIO}_3\times\frac{1\text{ mmol KIO}_3}{0.21400\text{ g KIO}_3}\times\frac{6\text{ mmol Na}_2\text{S}_2\text{O}_3}{\text{mmol KIO}_3}}{41.27\text{ mL Na}_2\text{S}_2\text{O}_3} = \underline{\underline{0.08411\text{ M Na}_2\text{S}_2\text{O}_3}}$$

11-25. total mmol NaOH = 25.00×0.00923 mmol/mL = 0.23075

mmol HCl added = 13.33 mL $\times$ 0.01007 mmol/mL = 0.13423

mmol NaOH consumed by analyte = 0.23075 - 0.13423 = 0.09652 mmol NaOH

$$\frac{0.0965\text{ mmol NaOH}\times\frac{1\text{ mmol H}_2\text{SO}_4}{2\text{ mmol NaOH}}\times\frac{1\text{ mmol S}}{1\text{ mmol H}_2\text{SO}_4}\times\frac{0.03207\text{ g S}}{\text{mmol S}}}{4.476\text{ g sample}}\times 10^6\text{ ppm} = \underline{\underline{345.8\text{ ppm S}}}$$

11-27. $\mathcal{M}_{\text{As}_2\text{O}_3} = 197.84\text{ g/mol}$

$$\begin{aligned}\text{amount Ag}^+\text{ required for Ag}_3\text{AsO}_4 &= \text{total mmol Ag}^+ - \text{no. mmol KSCN}\\ &= 40.00\text{ mL}\times 0.07891\text{ M} - 11.27\text{ mL}\times 0.1000\text{ M}\\ &= 2.0294\text{ mmol Ag}^+\end{aligned}$$

$$\frac{2.0294\text{ mmol Ag}^+\times\frac{1\text{ mmol As}}{3\text{ mmol Ag}}\times\frac{1\text{ mmol As}_2\text{O}_3}{2\text{ mmol As}}\times\frac{0.19784\text{ g As}_2\text{O}_3}{\text{mmol As}_2\text{O}_3}}{1.223\text{ g sample}}\times 100\% = \underline{\underline{5.471\%\text{ As}_2\text{O}_3}}$$

11-28. $$\frac{37.31\ \text{mL Hg}^{2+} \times 0.009372\frac{\text{mmol Hg}^{2+}}{\text{mL Hg}^{2+}} \times \frac{4\ \text{mmol analyte}}{\text{mmol Hg}^{2+}} \times 0.07612\frac{\text{g analyte}}{\text{mmol analyte}}}{1.455\ \text{g sample}} \times 100\% =$$

$$\underline{\underline{7.317\%\ (NH_2)_2CS}}$$

11-30. (a) $$\frac{0.1016\ \text{g}\ C_6H_5COOH \times \frac{1\ \text{mmol}\ C_6H_5COOH}{0.12212\ \text{g}\ C_6H_5COOH} \times \frac{1\ \text{mmol Ba(OH)}_2}{2\ \text{mmol}\ C_6H_5COOH}}{44.42\ \text{mL}} = \underline{\underline{9.36 \times 10^{-3}\ \text{M Ba(OH)}_2}}$$

(b) $$\left(\frac{\sigma_M}{M}\right)^2 = \left(\frac{\sigma_W}{W}\right)^2 + \left(\frac{\sigma_V}{V}\right)^2 = \left(\frac{0.2\ \text{mg}}{101.6\ \text{mg}}\right)^2 + \left(\frac{0.03\ \text{mL}}{44.42\ \text{mL}}\right)^2 = 4.33 \times 10^{-6}$$

$$\frac{\sigma_M}{M} = \sqrt{4.33 \times 10^{-6}} = 2.08 \times 10^{-3} = 2.1 \times 10^{-3}$$

$$\sigma_M = 2.08 \times 10^{-3} \times 9.36 \times 10^{-3} = \underline{\underline{1.9 \times 10^{-5} \text{M}}}$$

(c) Assuming that the systematic error in volume is zero

$$\frac{\Delta M}{M} = \frac{\Delta W}{W} = \frac{-0.3\ \text{mg}}{101.6\ \text{mg}} = -2.95 \times 10^{-3} \quad \text{or} \quad \underline{\underline{-3\ \text{ppt}}}$$

11-32. (a) $$0.3147\ \text{g}\ Na_2C_2O_4 \times \frac{1\ \text{mmol}\ Na_2C_2O_4}{134.0\ \text{g}} \times \frac{2\ \text{mol}\ KMnO_4}{5\ \text{mol}\ Na_2C_2O_4} \times \frac{10^3\ \text{mmol}}{\text{mol}} \times \frac{1}{31.67\ \text{mL}\ KMnO_4} =$$

$$0.02966\frac{\text{mmol}\ KMnO_4}{\text{mL}} = \underline{\underline{0.02966\ \text{M}}}$$

(b) $$\frac{26.75\ \text{mL}\ KMnO_4 \times 0.002966\frac{\text{mmol}}{\text{mL}} \times \frac{5\ \text{mmol}Fe_2O_3}{2\ \text{mmol}\ KMnO_4} \times \frac{0.15969\ \text{g}\ Fe_2O_3}{\text{mmol}\ Fe_2O_3}}{0.6656\ \text{g sample}} \times 100\% = 47.59\%\ Fe_2O_3$$

11-34. (a) $$c = \frac{10.12\ \text{g} \times \frac{1\ \text{mol}}{277.85\ \text{g}}}{2.000\ \text{L}} = \underline{\underline{1.821 \times 10^{-2}\ \text{M}}}$$

(b) $$[Mg^{2+}] = \underline{\underline{1.821 \times 10^{-2}\ \text{M}}}$$

(c) $[Cl^-] = 2[Mg^{2+}] + [K^+] = 3(1.821 \times 10^{-2}) = \underline{\underline{5.463 \times 10^{-2}\ M}}$

(d) $\dfrac{10.12\ g}{2.00\ L} \times \dfrac{1\ L}{1000\ mL} \times 100\% = \underline{\underline{0.506\%\ (w/v)}}$

(e) $\dfrac{1.821 \times 10^{-2}\ mmol\ KCl \cdot MgCl_2}{mL} \times \dfrac{3\ mmol\ Cl^-}{mmol\ KCl \cdot MgCl_2} \times 25.0\ mL = \underline{\underline{1.366\ mmol\ Cl^-}}$

(f) $1.821 \times 10^{-2} \dfrac{mmol\ salt}{mL} \times \dfrac{1\ mmol\ K^+}{mmol\ salt} \times \dfrac{39.10\ mg}{mmol\ K^+} \times \dfrac{1000\ mL}{L} = \dfrac{712\ mg\ K^+}{L} = \underline{\underline{712\ ppm\ K^+}}$

Chapter 12

12-1. **(a)** The initial pH of the NH_3 solution will be less than that for the solution containing NaOH. With the first addition of titrant, the pH of the NH_3 solution will decrease rapidly and then level off and become nearly constant throughout the middle part of the titration. In contrast, additions of standard acid to the NaOH solution will cause the pH of the NaOH solution to decrease gradually and nearly linearly until the equivalent point is approached. The equivalence point for the NH_3 solution will be well below seven, whereas for the NaOH solution it will be exactly seven.

(b) Beyond the equivalence point, the pH is determined by the excess titrant. Thus the curves become identical in this region.

12-3. The limited sensitivity of the eye to small color differences requires that there be a roughly tenfold excess of one or the other form of the indicator to be present in order for the color change to be seen. This change corresponds to a pH range of ± 1 pH unit about the pK of the indicator.

12-5. The standard reagents in neutralization titrations are always strong acids or strong bases because the reactions with this type of reagent are more complete than with those of their weaker counterparts. Sharper end points are the consequence of this difference.

12-7. The *buffer capacity* of a solution is the number of moles of hydronium ion or hydroxide ion needed to cause 1.00 L of the buffer to undergo a unit change in pH.

12-9. **(a)**

$$[H_3O^+] = K_a \times \frac{\text{mmol HOAc}}{\text{mmol OAc}^-} = K_a \times \frac{10.0}{8.00} = 2.50\,K_a$$

(b)

$$[H_3O^+] = K_a \times \frac{100 \times 0.175 - 100 \times 0.0500}{100 \times 0.0500} = K_a \times \frac{12.5}{5.00} = 2.50\,K_a$$

(c)

$$[H_3O^+] = K_a \times \frac{40.0 \times 0.1200}{160.0 \times 0.0420 - 40.0 \times 0.1200} = K_a \times \frac{4.80}{1.92} = 2.50\,K_a$$

The three solutions will have the same pH since the ratios of the amounts of weak acid to conjugate base are identical. They will differ, however, in buffer capacity with (a) having the greatest and (c) the least.

12-10. **(a)** The ideal system would have a pK_a of 3.5 or a K_a of 3.2×10^{-4}.

Malic acid/sodium hydrogen malate with K_a of 3.48×10^{-4} could be used.

(c) The ideal system would have a $pK_a = 9.3$ or $K_a = 5.0 \times 10^{-10}$.

NH_4Cl/NH_3 with $K_a = 5.7 \times 10^{-10}$.

12-11. The sharper end point will be observed with the solute having the larger K_b.

(a) For NaOCl,

$$K_b = \frac{1.00 \times 10^{-14}}{3.0 \times 10^{-8}} = 3.3 \times 10^{-7}$$

For $HONH_2$,

$$K_b = \frac{1.00 \times 10^{-14}}{1.1 \times 10^{-6}} = 9.1 \times 10^{-9}$$

Thus NaOCl

(c) For $HONH_2$, $K_b = 9.1 \times 10^{-9}$ (part a)

For CH_3NH_2,

$$K_b = \frac{1.00 \times 10^{-14}}{2.3 \times 10^{-11}} = 4.3 \times 10^{-4}$$

Thus CH_3NH_3

12-12. The sharper end point will be observed with the solute having the larger K_a.

(a) For HNO_2, $K_a = 7.1 \times 10^{-4}$

For HIO_3, $K_a = 1.7 \times 10^{-1}$ Thus HIO_3

(c) For HOCl, $K_a = 3.0 \times 10^{-8}$

For $CH_3COCOOH$, $K_a = 3.2 \times 10^{-3}$ Thus $CH_3COCOOH$

12-14. For methyl orange, $pK_a = 3.46 = -\log K_a$

$K_a = \text{antilog}(-3.46) = 3.47 \times 10^{-4}$

$$InH^+ + H_2O \rightleftarrows In + H_3O^+$$

$$\frac{[In][H_3O^+]}{[InH^+]} = 3.47 \times 10^{-4}$$

$$[H_3O^+] = \frac{3.47 \times 10^{-4}[InH^+]}{[In]} = 3.47 \times 10^{-4} \times 1.64 = 5.69 \times 10^{-4}$$

$$pH = -\log 5.69 \times 10^{-4} = \underline{\underline{3.24}}$$

12-16. (a) At $0°C$, $K_w = 1.14 \times 10^{-15}$ (see Problem 12 – 15)

$$pK_w = -\log 1.14 \times 10^{-15} = \underline{\underline{14.94}}$$

12-17. $[H_3O^+][OH^-] = K_w$ $\quad pH + pOH = pK_w$

(a) $pH = pK_w - pOH = 14.94 - 2.00 = \underline{\underline{12.94}}$ (see Solution 12-16)

12-18. $$\frac{14.0 \text{ g HCl}}{100 \text{ g soln}} \times \frac{1.054 \text{ g soln}}{\text{mL soln}} \times \frac{1 \text{ mmol HCl}}{0.03646 \text{ g HCl}} = 4.047 \text{ M}$$

$[H_3O^+] = 4.047$ M and $pH = -\log 4.047 = \underline{\underline{-0.607}}$

12-20. The solution is so dilute that we must take into account the contribution of water to $[OH^-]$, which is equal to $[H_3O^+]$. Thus,

$$[OH^-] = 2.0 \times 10^{-8} + [H_3O^+] = 2.0 \times 10^{-8} + \frac{1.00 \times 10^{-14}}{[OH^-]}$$

$$[OH^-]^2 - 2.0 \times 10^{-8}[OH^-] - 1.00 \times 10^{-14} = 0$$

$$[OH^-] = 1.105 \times 10^{-7}$$

$pOH = -\log 1.105 \times 10^{-7} = 6.957$ and $pH = 14.00 - 6.957 = \underline{\underline{7.04}}$

12-23. In each part, (20.0 mL HCl × 0.200 mmol HCl/mL) = 4.00 mmol HCl taken

(a) $c_{\mathrm{HCl}} = [\mathrm{H_3O^+}] = \dfrac{4.00\ \text{mmol HCl}}{(20.0+25.0)\ \text{mL soln}} = 0.0889\ \text{M}$

$\text{pH} = -\log 0.0889 = \underline{1.05}$

(b) Same as in part (a); $\text{pH} = \underline{1.05}$

(c) $c_{\mathrm{HCl}} = (4.00 - 25.0 \times 0.132)/(20.0 + 25.0) = 1.556 \times 10^{-2}\ \text{M}$

$[\mathrm{H_3O^+}] = 1.556 \times 10^{-2}\ \text{M}$ and $\text{pH} = -\log(1.556 \times 10^{-2}) = \underline{1.81}$

(d) As in part (c), $c_{\mathrm{HCl}} = 1.556 \times 10^{-2}$ and $\text{pH} = \underline{1.81}$

(The presence of NH_4^+ will not alter the pH significantly.)

(e) $c_{\mathrm{NaOH}} = (25.0 \times 0.232\ -\ 4.00)/(45.0) = 4.00 \times 10^{-2}$

$\text{pOH} = -\log 4.00 \times 10^{-2} = 1.398$ and $\text{pH} = 14.00 - 1.398 = \underline{12.60}$

12-24. **(a)** $[\mathrm{H_3O^+}] = 0.050$ and $\text{pH} = -\log 0.0500 = \underline{1.30}$

(b) $\mu = (0.0500 \times 1^2 + 0.0500 \times 1^2)/2 = 0.0500$

$\gamma_{\mathrm{H_3O^+}} = 0.85 \qquad (\text{Table } 9-1)$

$a_{\mathrm{H_3O^+}} = 0.85 \times 0.0500 = 0.043$

$\text{pH} = -\log 0.043 = \underline{1.37}$

12-26. $\dfrac{[\mathrm{H_3O^+}][\mathrm{OCl^-}]}{[\mathrm{HOCl}]} = K_a = 3.0 \times 10^{-8}$

$[\mathrm{H_3O^+}] = [\mathrm{OCl^-}]$ and $[\mathrm{HOCl}] = c_{\mathrm{HOCl}} - [\mathrm{H_3O^+}] \approx c_{\mathrm{HOCl}}$

$[\mathrm{H_3O^+}]^2 / c_{\mathrm{HOCl}} = 3.0 \times 10^{-8}$

$[\mathrm{H_3O^+}] = \sqrt{c_{\mathrm{HOCl}} \times 3.0 \times 10^{-8}}$

Thus when c_{HOCl} =

(a) 0.100 $[H_3O^+] = 5.477 \times 10^{-5}$ and pH = $\underline{4.26}$

(b) 0.0100 $[H_3O^+] = 1.732 \times 10^{-5}$ and pH = $\underline{4.76}$

(c) 1.00×10^{-4} $[H_3O^+] = 1.732 \times 10^{-6}$ and pH = $\underline{5.76}$

12-28. $NH_3 + H_2O \rightleftarrows NH_4^+ + OH^-$ $\quad K_b = \dfrac{1.00 \times 10^{-14}}{5.70 \times 10^{-10}} = 1.75 \times 10^{-5}$

(a) $$\frac{[NH_4^+][OH^-]}{[NH_3]} = \frac{[OH^-]}{0.100 - [OH^-]} = \frac{[OH^-]^2}{0.100} = 1.75 \times 10^{-5}$$

$$[OH^-] = \sqrt{1.75 \times 10^{-5} \times 0.100} = 1.323 \times 10^{-3}$$

$$pH = 14.00 - (-\log 1.323 \times 10^{-3}) = \underline{\underline{11.12}}$$

(b) Proceeding in the same way, we obtain pH = $\underline{\underline{10.62}}$

(c) Proceeding in the same way, we obtain

$$[OH^-] = 4.19 \times 10^{-5} \quad \text{and} \quad pH = 9.62$$

To obtain a more exact solution, we write

$$\frac{[OH^-]^2}{1.00 \times 10^{-4} - [OH^-]} = 1.75 \times 10^{-5}$$

$$[OH^-]^2 + 1.75 \times 10^{-5}[OH^-] - 1.75 \times 10^{-9} = 0$$

$$[OH^-] = 3.40 \times 10^{-5}$$

$$pH = 14.00 - (-\log 3.40 \times 10^{-5}) = \underline{\underline{9.53}}$$

12-30. $C_5H_{11}N + H_2O \rightleftarrows C_5H_{11}NH^+ + OH^-$ $\quad K_b = \dfrac{1.00 \times 10^{-14}}{7.50 \times 10^{-12}} = 1.333 \times 10^{-3}$

(a) Because K_b is large we shall not make the usual approximations and will write

$$\frac{[OH^-]^2}{0.100 - [OH^-]} = 1.333 \times 10^{-3}$$

$$[OH^-]^2 + 1.333 \times 10^{-3}[OH^-] - 1.333 \times 10^{-4} = 0$$

$$[OH^-] = 1.090 \times 10^{-2}$$

$$pH = 14.00 - (-\log 1.00 \times 10^{-2}) = \underline{\underline{12.04}}$$

Proceeding in the same way, we obtain

(b) $[OH^-] = 3.045 \times 10^{-3}$ and $pH = \underline{\underline{11.48}}$

(c) $[OH^-] = 9.34 \times 10^{-5}$ and $pH = \underline{\underline{9.97}}$

If the approximate method is used, the pH values are

(a) 12.06 **(b)** 11.56 **(c)** 10.56

12-32. (a) $\mathcal{M}_{HA} = 90.079$

$$c_{HA} = 43.0 \text{ g HA} \times \frac{1 \text{ mmol HA}}{0.090079 \text{ g HA}} \times \frac{1}{500 \text{ mL soln}} = 0.9547 \text{ M HA}$$

$$\frac{[H_3O^+][A^-]}{[HA]} = 1.38 \times 10^{-4}$$

$$[H_3O^+] = [A^-] \quad \text{and} \quad [HA] = 0.9547 - [H_3O^+] = 0.9547$$

$$[H_3O^+]^2/0.9547 = 1.38 \times 10^{-4}$$

$$[H_3O^+] = \sqrt{1.38 \times 10^{-4} \times 0.9547} = 1.148 \times 10^{-2}$$

$$pH = -\log 1.148 \times 10^{-2} = \underline{\underline{1.94}}$$

(b) $c_{HA} = 0.9547 \times 25/250 = 0.09547 \text{ M}$

Proceeding as in part (a), we obtain

$$[H_3O^+] = 3.63 \times 10^{-3} \text{ and pH} = \underline{\underline{2.44}}$$

To obtain a more exact answer,

$$[H_3O^+] = [A^-] \text{ and } [HA] = 0.09547 - [H_3O^+]$$

$$\frac{[H_3O^+]}{0.09547 - [H_3O^+]} = 1.38 \times 10^{-4}$$

$$[H_3O^+]^2 + 1.38 \times 10^{-4}[H_3O^+] - 1.317 \times 10^{-5} = 0$$

$$[H_3O^+] = 3.561 \times 10^{-3} \quad \text{and} \quad \text{pH} = \underline{\underline{2.45}}$$

(c) $c_{HA} = 0.09547 \times 10/1000 = 9.547 \times 10^{-4}$

Here we cannot assume that $[H_3O^+] << 9.547 \times 10^{-4}$. Thus,

$$[H_3O^+] = [A^-] \text{ and } [HA] = 9.547 \times 10^{-4} - [H_3O^+]$$

$$\frac{[H_3O^+]^2}{9.547 \times 10^{-4} - [H_3O^+]} = 1.38 \times 10^{-4}$$

$$[H_3O^+]^2 + 1.38 \times 10^{-4}[H_3O^+] - 1.317 \times 10^{-7} = 0$$

$$[H_3O^+] = 3.004 \times 10^{-4} \text{ and pH} = -\log 3.004 \times 10^{-4} = \underline{\underline{3.52}}$$

12-34. For formic acid, HA $K_{HA} = [H_3O^+][A^-]/[HA] = 1.80 \times 10^{-4}$

Throughout Solution 12-34

$$\text{amount HA taken} = 20.00 \text{ mL} \times \frac{0.200 \text{ mmol}}{\text{mL}} = 4.00 \text{ mmol}$$

(a) $c_{HA} = 4.00/45.0 = 8.89 \times 10^{-2}$

$$[H_3O^+] = [A^-] \text{ and } [HA] = 8.89 \times 10^{-2} - [H_3O^+]$$

$$[H_3O^+]^2/(8.89\times10^{-2}-[H_3O^+]) = 1.80\times10^{-4}$$

$$[H_3O^+]^2+1.80\times10^{-4}[H_3O^+]-8.89\times10^{-2}\times1.80\times10^{-4} = 0$$

$[H_3O^+] = 3.91\times10^{-3}$ and pH = $\underline{\underline{2.41}}$

(b) Amount NaOH added = 25.0×0.160 = 4.00 mmol.

Therefore, we have a solution of NaA.

$$c_{NaA} = 4.00/(20.0 + 25.0) = 8.89\times10^{-2}$$

$$A^- + H_2O \rightleftarrows HA + OH^-$$

$$K_b = 1.00\times10^{-14}/1.80\times10^{-4} = 5.56\times10^{-11}$$

$[OH^-]$ = [HA] and $[A^-] = 8.89\times10^{-2} - [OH^-] = 8.89\times10^{-2}$

$$[OH^-]^2/(8.89\times10^{-2}) = 5.56\times10^{-11}$$

$$[OH^-] = \sqrt{8.89\times10^{-2}\times5.56\times10^{-11}} = 2.22\times10^{-6}$$

$$pH = 14.00 - (-\log 2.22\times10^{-6}) = \underline{\underline{8.35}}$$

(c) Amount NaOH added = 25.0×0.200 = 5.00 mmol

Therefore, we have an excess of NaOH and the pH is determined by its concentration.

$$[OH^-] = c_{NaOH} = (5.00-4.00)/45.0 = 2.222\times10^{-2}$$

$$pH = 14.00-(-\log 2.222\times10^{-2}) = \underline{\underline{12.35}}$$

(d) Amount NaA added = 25.0×0.200 = 5.00 mmol

$$c_{NA} = 5.00/45.00 = 0.1111\text{ M} \approx [A^-]$$

$$c_{HA} = 4.00/45.00 = 0.0889\text{ M} \approx [HA]$$

$$[H_3O^+]\,0.1111/0.0889 = 1.80\times10^{-4}$$

$$[H_3O^+] = 1.440\times10^{-4} \text{ and } pH = \underline{\underline{3.84}}$$

12-37. In each part of this problem, a buffer mixture of a weak acid HA and its conjugate base NaA is formed and Equations 12-6 and 12-7 (page 277) apply. That is,

$$[HA] = c_{HA} - [H_3O^+] + [OH^-]$$

$$[A^-] = c_{NaA} + [H_3O^+] - [OH^-]$$

In each case, we will assume initially that $[H_3O^+]$ and $[OH^-]$ are much smaller than the molar concentration of the acid and conjugate so that $[A^-] \approx c_{NaA}$ and $[HA] \approx c_{HA}$. These assumptions then lead to Equation 12-10.

$$[H_3O^+] = K_a c_{HA}/c_{NaA}$$

(a)

$$c_{HA} = 9.20\text{ g HA}\times\frac{1\text{ mol HA}}{90.08\text{ g HA}}\times\frac{1}{1.00\text{ L}} = 0.1021\text{ M}$$

$$c_{NaA} = 11.15\text{ g NaA}\times\frac{1\text{ mol NaA}}{112.06\text{ g NaA}}\times\frac{1}{1.00\text{ L}} = 0.0995\text{ M}$$

$$[H_3O^+] = 1.38\times10^{-4}\times0.1021/0.0995 = 1.416\times10^{-4}$$

Note that $[H_3O^+]$ (and $[OH^-]$) $<< c_{HA}\, c_{NaA}$ as assumed. Therefore,

$$pH = -\log 1.416\times10^{-4} = \underline{\underline{3.85}}$$

(b) $c_{HA} = 0.0550$ and $c_A = 0.0110$

$$[H_3O^+] = 1.75\times10^{-5}\times0.0550/0.0110 = 8.75\times10^{-5} \quad\text{and}\quad pH = \underline{\underline{4.06}}$$

(c)

$$\text{Original amount HA} = 3.00\text{ g}\times\frac{\text{mmol HA}}{0.13812\text{ g}} = 21.72\text{ mmol}$$

$$\text{Original amount NaOH} = 50.0 \text{ mL} \times \frac{0.1130 \text{ mmol}}{\text{mL}} = 5.65 \text{ mmol}$$

$$c_{HA} = (21.72 - 5.65)/500 = 3.214 \times 10^{-2}$$

$$c_{NaA} = 5.65/500 = 1.130 \times 10^{-2}$$

$$[H_3O^+] = 1.06 \times 10^{-3} \times 3.214 \times 10^{-2}/1.130 \times 10^{-2} = 3.015 \times 10^{-3}$$

$$pH = 2.521$$

Note, however, that $[H_3O^+]$ is *not* $<< c_{HA}$ and c_{NaA}.

Thus, Equations 12-6 and 12-7 must be used.

$$[HA] = 3.214 \times 10^{-2} - [H_3O^+] + \cancel{[OH^-]}$$

$$[A^-] = 1.130 \times 10^{-2} + [H_3O^+] - \cancel{[OH^-]}$$

Certainly, $[OH^-]$ will be negligible since the solution is acidic.

Substituting into the dissociation-constant expression gives

$$\frac{[H_3O^+](1.130 \times 10^{-2} + [H_3O^+])}{3.214 \times 10^{-2} - [H_3O^+]} = 1.06 \times 10^{-3}$$

Rearranging gives

$$[H_3O^+]^2 + 1.236 \times 10^{-2}[H_3O^+] - 3.407 \times 10^{-5} = 0$$

$$[H_3O^+] = 2.321 \times 10^{-3} \quad \text{and} \quad pH = \underline{\underline{2.63}}$$

(d) Here, Equation 12-10 gives $[H_3O^+] = 4.3 \times 10^{-2}$, which is clearly not $<< c_{HA}$ or c_{NaH} and we must proceed as in part (c).

This leads to

$$\frac{[H_3O^+](0.100 + [H_3O^+])}{0.0100 - [H_3O^+]} = 4.3 \times 10^{-1}$$

$$[H_3O^+]^2 + 0.53[H_3O^+] - 4.3 \times 10^{-3} = 0$$

$[H_3O^+] = 7.99 \times 10^{-3}$ and $pH = \underline{\underline{2.10}}$

12-39. (a) $\Delta pH = \underline{\underline{0.00}}$

(c) pH diluted solution $= 14.000 - (-\log 0.00500) = 11.699$

pH undiluted solution $= 14.000 - (-\log 0.0500) = 12.699$

$\Delta pH = \underline{\underline{-1.000}}$

(e) $OAc^- + H_2O \rightleftarrows HOAc + OH^-$

$$\frac{[HOAc][OH^-]}{[OAc^-]} = \frac{1.00 \times 10^{-14}}{1.75 \times 10^{-5}} = 5.71 \times 10^{-10} = K_b$$

Here we can use an approximation solution because K_b is so very small. For the undiluted sample

$$\frac{[OH^-]^2}{0.0500} = 5.71 \times 10^{-10}$$

$$[OH^-] = \sqrt{5.71 \times 10^{-10} \times 0.0500} = 5.343 \times 10^{-6}$$

$$pH = 14.00 - (-\log 5.345 \times 10^{-6}) = 8.728$$

For the diluted sample

$$[OH^-] = \sqrt{5.71 \times 10^{-10} \times 0.00500} = 1.690 \times 10^{-6}$$

$$pH = 14.00 - (-\log 1.690 \times 10^{-6}) = 8.228$$

$$\Delta pH = 8.228 - 8.728 = \underline{\underline{-0.500}}$$

(g) For the undiluted solution, we apply Equations 12-6 and 12-7 and obtain

$$[HOAc] = 0.500 - [H_3O^+] + \cancel{[OH^-]}$$

$$[OAc^-] = 0.500 + [H_3O^+] - \cancel{[OH^-]}$$

$$K_a = 1.75 \times 10^{-5} = \frac{[H_3O^+](0.500 + [H_3O^+])}{0.500 - [H_3O^+]}$$

$$8.75 \times 10^{-6} - 1.75 \times 10^{-5}[H_3O^+] = 0.500[H_3O^+] + [H_3O^+]^2$$

$$0 = [H_3O^+]^2 + (0.500 + 1.75 \times 10^{-5})[H_3O^+] - 8.75 \times 10^{-5}$$

Solving this quadratic equation gives

$$[H_3O^+] = 1.74988 \times 10^{-5}$$

$$pH = 4.75699 = 4.757$$

Repeating these calculations for 0.0500 M solution gives

$$[H_3O^+] = 1.748776 \times 10^{-5}$$

$$pH = 4.757266 = 4.757$$

$$\Delta pH = \underline{\underline{0.000}}$$

12-42. For lactic acid, $K_a = 1.38 \times 10^{-4} = [H_3O^+][A^-]/[HA]$

Throughout this problem we will base calculations of Equations 12-6 and 12-7. That is,

$$[A^-] = c_{NaA} + [H_3O^+] - \cancel{[OH^-]}$$

$$[HA] = c_{HA} - [H_3O^+] + \cancel{[OH^-]}$$

$$\frac{[H_3O^+](c_{NaA} + [H_3O^+])}{c_{HA} - [H_3O^+]} = 1.38 \times 10^{-4}$$

This equation rearranges to

$$[H_3O^+]^2 + (1.38 \times 10^{-4} + c_{NaA})[H_3O^+] - 1.38 \times 10^{-4} c_{HA} = 0$$

(b) Before addition of acid

$$[H_3O^+]^2 + (1.38 \times 10^{-4} + 0.0200)[H_3O^+] - 1.38 \times 10^{-4} \times 0.0800 = 0$$

$$[H_3O^+] = 5.341 \times 10^{-4} \quad \text{and} \quad pH = 3.272$$

After adding acid

$$c_{HA} = (100 \times 0.0800 + 0.500)/100 = 0.085\ M$$

$$c_{NaA} = (100 \times 0.0200 - 0.500)/100 = 0.015\ M$$

$$[H_3O^+]^2 + (1.38 \times 10^{-4} + 0.015) - 1.38 \times 10^{-4} \times 0.085 = 0$$

$$[H_3O^+] = 7.388 \times 10^{-4} \quad \text{and} \quad pH = 3.131$$

$$\Delta pH = 3.131 - 3.272 = \underline{\underline{-0.141}}$$

12-43. $pH = 3.50$ and $[H_3O^+] = \text{antilog}(-3.50) = 3.162 \times 10^{-4}$

$$[H_3O^+][A^-]/[HA] = 1.80 \times 10^{-4}$$

$$[A^-]/[HA] = 1.80 \times 10^{-4}/3.162 \times 10^{-4} = 0.5693$$

$$[HA] = 1.00 \quad \text{and} \quad [A^-] = 0.5693 \times 1.00 = 0.5693\ M$$

$$\text{mass HCOONa} = 0.5693 \frac{\text{mmol NaA}}{\text{mL}} \times 400\ \text{mL} \times \frac{0.06801\ \text{g NaA}}{\text{mmol NaA}}$$

$$= \underline{\underline{15.5\ \text{g sodium formate}}}$$

12-45. $pH = 3.37$ and $[H_3O^+] = \text{antilog}(-3.37) = 4.266 \times 10^{-4}$

$$4.0 \times 10^{-4} = [H_3O^+][A^-]/[HA] = 4.266 \times 10^{-4}[A^-]/[HA]$$

$[A^-]/[HA] \quad = \quad 0.9377$

Let V_{HCl} = mL 0.200 M HCl added

amount HA formed = amount HCl added = $V_{HCl} \times 0.200$ mmol

amount NaA remaining = original amout NaA - amount V_{HCl} added

= 250×0.300 - $V_{HCl} \times 0.200$) mmol

Total volume of solution = $(250 + V_{HCl}$) mL

$$c_{HA} \quad = \quad 0.200\ V_{HCl}/(250 + V_{HCl}) \quad \approx \quad [HA]$$

$$c_{NaA} \quad = \quad (75.0 - 0.200\ V_{HCl})/(250 + V_{HCl}) \quad \approx \quad [A^-]$$

Substituting

$$\frac{[A^-]}{[HA]} \quad = \quad \frac{(75.0 - 0.200\ V_{HCl})/(250 + V_{HCl})}{0.200\ V_{HCl}/(250 + V_{HCl})} \quad = \quad 0.9377$$

$$75.0 - 0.200\ V_{HCl} \quad = \quad 0.200\ V_{HCl} \times 0.9377 \quad = \quad 0.1875\ V_{HCl}$$

$$V_{HCl} \quad = \quad 75.0/(0.200 + 0.1875) \quad = \quad \underline{\underline{194\text{ mL HCl}}}$$

12-47. Before the equivalence point (50.00 mL), we calculate the number of mmol NaOH remaining from the original number of mmol NaOH present minus the number of mmol of HCl added. The $[OH^-]$ is then obtained from the number of mmol of NaOH remaining divided by the total solution volume. The $[OH^-]$ is then used to obtain pOH and from this the pH. At the equivalence point the pOH and pH are obtained from $\sqrt{K_w}$. After the equivalence point, we calculate the excess HCl from the number of mmol of HCl added minus the number of mmol of NaOH originally present. The number of mmol of HCl divided by the total volume gives the H_3O^+ concentration and from this the pH. The spreadsheet shows the resulting pH values after each increment of NaOH. The plot is an XY (Scatter) plot of pH (17:116) vs. Vol. HCl (A7:A16).

12-47.

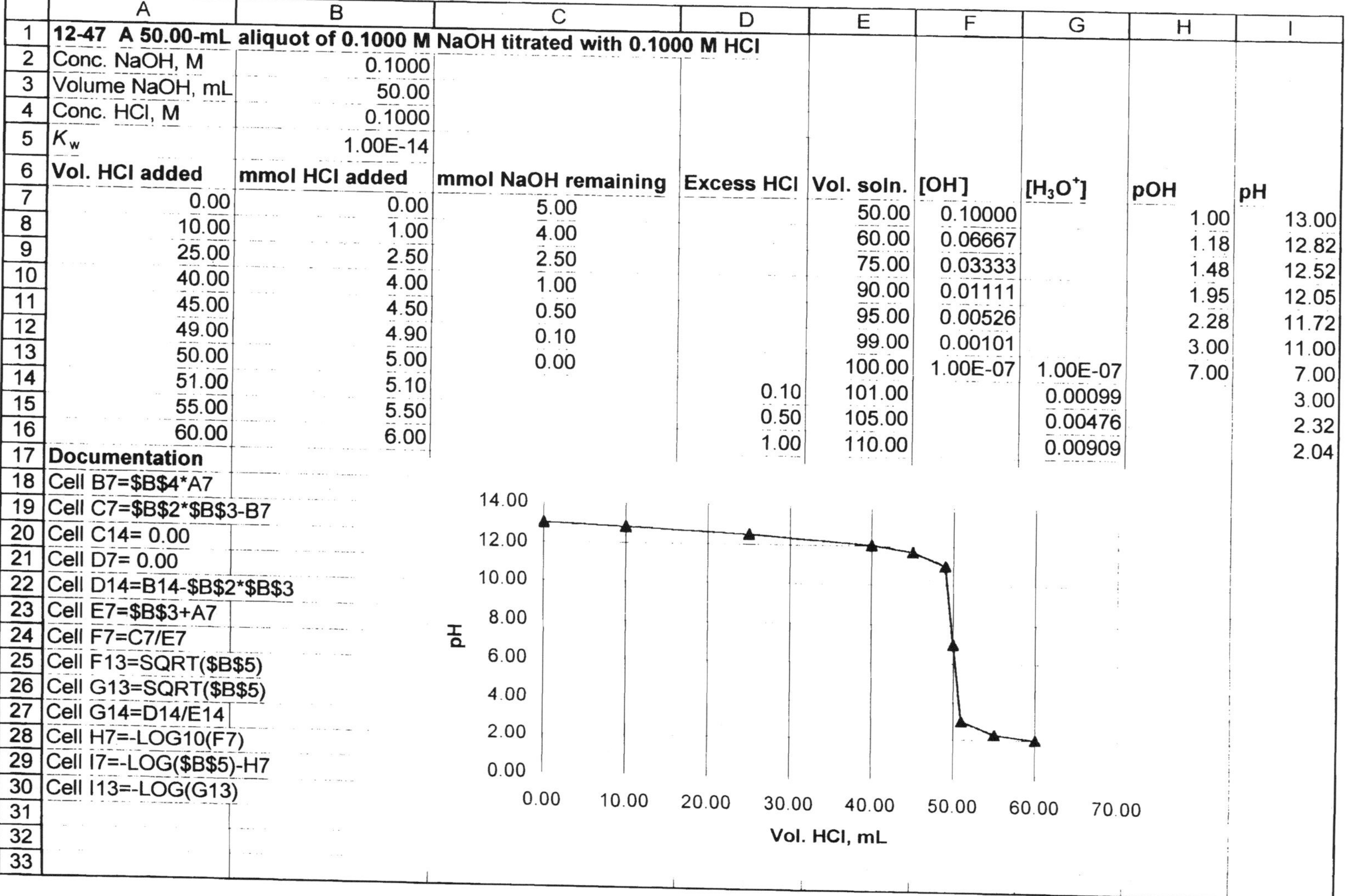

	A	B	C	D	E	F	G	H	I
1	**12-47**	**A 50.00-mL aliquot of 0.1000 M NaOH titrated with 0.1000 M HCl**							
2	Conc. NaOH, M	0.1000							
3	Volume NaOH, mL	50.00							
4	Conc. HCl, M	0.1000							
5	K_w	1.00E-14							
6	**Vol. HCl added**	**mmol HCl added**	**mmol NaOH remaining**	**Excess HCl**	**Vol. soln.**	**[OH^-]**	**[H_3O^+]**	**pOH**	**pH**
7	0.00	0.00	5.00		50.00	0.10000		1.00	13.00
8	10.00	1.00	4.00		60.00	0.06667		1.18	12.82
9	25.00	2.50	2.50		75.00	0.03333		1.48	12.52
10	40.00	4.00	1.00		90.00	0.01111		1.95	12.05
11	45.00	4.50	0.50		95.00	0.00526		2.28	11.72
12	49.00	4.90	0.10		99.00	0.00101		3.00	11.00
13	50.00	5.00	0.00		100.00	1.00E-07	1.00E-07	7.00	7.00
14	51.00	5.10		0.10	101.00		0.00099		3.00
15	55.00	5.50		0.50	105.00		0.00476		2.32
16	60.00	6.00		1.00	110.00		0.00909		2.04
17	**Documentation**								
18	Cell B7=$B$4*A7								
19	Cell C7=$B$2*$B$3-B7								
20	Cell C14= 0.00								
21	Cell D7= 0.00								
22	Cell D14=B14-$B$2*$B$3								
23	Cell E7=$B$3+A7								
24	Cell F7=C7/E7								
25	Cell F13=SQRT($B$5)								
26	Cell G13=SQRT($B$5)								
27	Cell G14=D14/E14								
28	Cell H7=-LOG10(F7)								
29	Cell I7=-LOG($B$5)-H7								
30	Cell I13=-LOG(G13)								
31									
32									
33									

12-48. Let us calculate pH when 24.95 and 25.05 mL of reagent have been added.

24.95 mL reagent

$$c_{A^-} \approx \frac{\text{amount KOH added}}{\text{total volume solution}} = \frac{24.95 \times 0.1000 \text{ mmol KOH}}{74.95 \text{ mL solution}} = \frac{2.495}{74.95}$$

$$c_{HA} \approx [HA] = \frac{\text{original amount HA} - \text{amount KOH added}}{\text{total volume solution}}$$

$$= \frac{(50.00 \times 0.0500 - 24.95 \times 0.1000) \text{ mmol HA}}{74.95 \text{ mL solution}}$$

$$= \frac{2.500 - 2.495}{74.95} = \frac{0.05}{74.95}$$

Substituting into Equation 12-10

$$[H_3O^+] = K_a \frac{c_{HA}}{c_{A^-}} = \frac{1.80 \times 10^{-4} \times 0.005/74.95}{2.495/74.95} = 3.607 \times 10^{-7}$$

$$\text{pH} = -\log 3.607 \times 10^{-7} = \underline{\underline{6.44}}$$

25.05 mL KOH

$$c_{KOH} = \frac{\text{amount KOH added} - \text{initial amount HA}}{(25.05 + 50.00) \text{ mL}}$$

$$= \frac{25.05 \times 0.1000 - 50.00 \times 0.0500}{75.05} = 6.66 \times 10^{-5} = [OH^-]$$

$$\text{pH} = 14.00 - (-\log 6.66 \times 10^{-5}) = \underline{\underline{9.82}}$$

Thus, the indicator should change color in the range of pH <u>6.5 to 9.8</u>. Cresol purple (range: 7.6 to 9.2) (Table 12-1) should be quite suitable.

12-50. (a) 0.00 mL NaOH

$c_{HA} = 0.1000\text{ M} \qquad [H_3O^+] = [A^-] \qquad [HA] = 0.100 - [H_3O^+]$

$$K_a = 7.1\times10^{-4} = \frac{[H_3O^+][A^-]}{[HA]} = \frac{[H_3O^+]^2}{0.1000-[H_3O^+]}$$

$$[H_3O^+]^2 + 7.1\times10^{-4}[H_3O^+] - 0.1000\times7.1\times10^{-4} = 0$$

The result is shown in cell E7 of the spreadsheet for this problem and pH = 2.09

5.00 mL NaOH

$$c_{NaA} = (5.00\times0.1000)/(50.00+5.00) = 9.09\times10^{-3}\text{ M}$$

$$c_{HA} = (50.00\times0.1000 - 5.00\times0.1000)/(55.00) = 8.18\times10^{-2}\text{ M}$$

We now have a buffer, but simplifying assumptions are not valid. Therefore, we must use Equations 12-6 and 12-7.

$$K_a = \frac{[H_3O^+](c_{NaA} + [H_3O^+] - [OH^-])}{(c_{HA} - [H_3O^+] + [OH^-])}$$

which arranges to

$$[H_3O^+]^2 + (K_a + c_{NaA})[H_3O^+] - K_a c_{HA} = 0 \qquad (1)$$

The solution is shown in cell E8 and pH = 2.38

The other preequivalence point calculations are performed by substitution into equation (1). The results are shown in the spreadsheet that follows.

50.00 mL NaOH

Here we are at the equivalence point and have a solution of NaA.

$$c_{NaA} = \frac{50.00\times0.1000}{50.00+50.00} = 0.05000\text{ M}$$

$$A^- + H_2O \rightleftarrows HA + OH^-$$

$$K_b = \frac{[OH^-][HA]}{[A^-]} = \frac{1.00 \times 10^{-14}}{7.1 \times 10^{-4}} = 1.408 \times 10^{-11}$$

$[HA] = [OH^-]$ and $[A^-] = 0.0500 - [OH^-] \approx 0.0500$

$$[OH^-]^2/0.05000 = 1.408 \times 10^{-11}$$

$$[OH^-] = \sqrt{0.005000 \times 1.408 \times 10^{-11}} = 8.390 \times 10^{-7}$$

$$[H_3O^+] = K_w/[OH^-]$$

The results are in cell E14, and pH = 7.92

51.00 mL NaOH

We now have an excess of NaOH.

$$[OH^-] \approx c_{NaOH} = (51.00 \times 0.1000 - 50.00 \times 0.1000)/101.0$$

$$= 9.901 \times 10^{-4}$$

$[H_3O^+] = K_w/[OH^-]$ (cell E15) and pH = 11.00

The other post-equivalence point data are treated in the same way.

(c) Here the acid is $C_5H_5NH^+$.

$$C_5H_5NH^+ + H_2O \rightleftarrows C_5HN_5 + 4H_3O^+$$

The calculations for this titration can be performed exactly as those in part (a). The results are shown in the spreadsheet that follows.

	A	B	C	D	E	F	G	H	I	J
1	**12-50 50.00 mL 0.1000 M acid with 0.1000 M NaOH**									
2	K_a for HNO_2	7.10E-04	Initial conc. acid	0.1000	K_w	1.00E-14				
3	K_a for HL	1.38E-04	Initial conc. NaOH	0.1000						
4	K_a for PyH^+	5.90E-06			(a)		(b)		(c)	
5	Vol. acid, mL	50.00			For nitrous acid		For lactic acid		For pyridinium chloride	
6	**Vol. NaOH, mL**	c_{acid}	$c_{conj.base}$	c_{NaOH}	**$[H_3O^+]$**	**pH**	**$[H_3O^+]$**	**pH**	**$[H_3O^+]$**	**pH**
7	0.00	0.10000	0.00000	0.0000	8.08E-03	2.09	3.65E-03	2.44	7.65E-04	3.12
8	5.00	0.08182	0.00909	0.0000	4.16E-03	2.38	1.09E-03	2.96	5.28E-05	4.28
9	15.00	0.05385	0.02308	0.0000	1.51E-03	2.82	3.16E-04	3.50	1.38E-05	4.86
10	25.00	0.03333	0.03333	0.0000	6.82E-04	3.17	1.37E-04	3.86	5.90E-06	5.23
11	40.00	0.01111	0.04444	0.0000	1.74E-04	3.76	3.44E-05	4.46	1.47E-06	5.83
12	45.00	0.00526	0.04737	0.0000	7.76E-05	4.11	1.53E-05	4.82	6.55E-07	6.18
13	49.00	0.00101	0.04949	0.0000	1.43E-05	4.85	2.81E-06	5.55	1.20E-07	6.92
14	50.00	0.00000	0.05000	0.0000	1.19E-08	7.92	5.25E-09	8.28	1.09E-09	8.96
15	51.00			0.0010	1.01E-11	11.00	1.01E-11	11.00	1.01E-11	11.00
16	55.00			0.0048	2.10E-12	11.68	2.10E-12	11.68	2.10E-12	11.68
17	60.00			0.0091	1.10E-12	11.96	1.10E-12	11.96	1.10E-12	11.96
18	**Spreadsheet Documentation**									
19	Applicable to all					Note that although the quadratic formula is not necessary for part(c),				
20	Cell B7=($D$2*$B$5-$D$3*A7)/($B$5+A7)					once it is entered in parts (a) and (b) it is just as easy to use it for				
21	Cell C7=$D$3*A7/($B$5+A7)					part (c).				
22	Cell D7=0									
23	Cell D15=(A15*$D$3-$B$5*$D$2)/($B$5+A15)									
24	Cells E15, G15, and I15=$F$2/D15									
25	For nitrous acid									
26	Cell E7=(-$B$2+SQRT($B$2^2-4*(-0.1*$B$2)))/2									
27	Cell F7=-LOG10(E7)									
28	Cell E8=(-($B$2+C8)+SQRT(($B$2+C8)^2-4*(-$B$2*B8)))/2									
29	Cell E14=$F$2/SQRT(C14*$F$2/$B$2)									
30	For lactic acid				For pyridinium chloride					
31	Cell G7=(-$B$3+SQRT($B$3^2-4*(-0.1*$B$3)))/2				Cell I7=(-$B$4+SQRT($B$4^2-4*(-0.1*$B$4)))/2					
32	Cell H7=-LOG10(G7)				Cell J7=-LOG10(I7)					
33	Cell G8=(-($B$3+C8)+SQRT(($B$3+C8)^2-4*(-$B$3*B8)))/2				Cell I8 =(-($B$4+C8)+SQRT(($B$4+C8)^2-4*(-$B$4*B8)))/2					
34	Cell G14=$F$2/SQRT(C14*$F$2/$B$3)				Cell I14=$F$2/SQRT(C14*$F$2/$B$4)					

12-51. (a) $NH_4^+ + H_2O \rightleftarrows H_3O^+ + NH_3 \qquad K_a = 5.70 \times 10^{-10}$

$NH_3 + H_2O \rightleftarrows NH_4^+ + OH^-$

$$K_b = \frac{K_w}{K_a} = 1.75 \times 10^{-5} = [OH^-][NH_4^+]/[NH_3]$$

0.00 mL 0.1000 M HCl

$[NH_4^+] = [OH^-] \qquad [NH_3] = 0.1000 - [OH^-] = 0.1000$

$[OH^-] = \sqrt{c_{NH_3} K_b} = \sqrt{0.1000 \times 1.75 \times 10^{-5}} = 1.323 \times 10^{-3}$

$[H_3O^+] = K_w/[OH^-]$ (cell E7), $\quad pH = 11.12$

5.00 mL 0.1000 M HCl

$c_{NH_4^+} = 5.00 \times 0.1000/(50.00 + 5.00) = 9.091 \times 10^{-3}$

$c_{NH_3} = (50.00 \times 0.1000 - 5.00 \times 0.1000)/55.00$

$= 8.182 \times 10^{-2}$

$[NH_3] \approx c_{NH_3} \qquad [NH_4^+] \approx c_{NH_4^+}$

Substituting these equalities into the expression for K_b gives upon rearranging

$[H_3O^+] = K_a c_{NH_4^+}/c_{NH_3}$ (cell E8), $\quad pH = 10.20$

The other preequivalence point data are treated in the same way. The answers appear in the spreadsheet that follows.

50.00 mL 0.1000 M HCl. This is the equivalence point at which

$c_{NH_4^+} = 50.00 \times 0.1000/100.0 = 0.05000$

$[H_3O^+] = [NH_3]$ and $[NH_4^+] = 0.05000 - [H_3O^+] \approx 0.05000$

$$[H_3O^+]^2/0.05000 = 5.70 \times 10^{-10}$$

$$[H_3O^+] = \sqrt{K_a c_{NH_4^+}}$$

The results are in cell E14, and pH = 5.27

51.00 mL HCl

$$c_{HCl} = [H_3O^+] = (51.00 \times 0.1000 - 50.00 \times 0.1000)/101.0$$

$$= 9.901 \times 10^{-4}$$

Other post-equivalence point data are obtained in the same way.

	A	B	C	D	E	F	G	H	I	J
1	**12-51 50.00 mL 0.1000 M base with 0.1000 M HCl**									
2	K_a for ammonium	5.70E-10	K_b for ammonia	1.75E-05	Initial conc. base	0.1000	K_w	1.00E-14		
3	K_a for hydrazinium	1.05E-08	K_b for hydrazine	9.52E-07	Initial conc. HCl	0.1000				
4	K_a for HCN	6.20E-10	K_b for cyanide	1.61E-05	(a)		(b)		(c)	
5	Vol. base, mL	50.00			For ammonia		For hydrazine		For sodium cyanide	
6	**Vol. HCl, mL**	c_{base}	$c_{conj.acid}$	c_{HCl}	**$[H_3O^+]$**	**pH**	**$[H_3O^+]$**	**pH**	**$[H_3O^+]$**	**pH**
7	0.00	0.10000	0.00000	0.0000	7.55E-12	11.12	3.24E-11	10.49	7.87E-12	11.10
8	5.00	0.08182	0.00909	0.0000	6.33E-11	10.20	1.17E-09	8.93	6.89E-11	10.16
9	15.00	0.05385	0.02308	0.0000	2.44E-10	9.61	4.50E-09	8.35	2.66E-10	9.58
10	25.00	0.03333	0.03333	0.0000	5.70E-10	9.24	1.05E-08	7.98	6.20E-10	9.21
11	40.00	0.01111	0.04444	0.0000	2.28E-09	8.64	4.20E-08	7.38	2.48E-09	8.61
12	45.00	0.00526	0.04737	0.0000	5.13E-09	8.29	9.45E-08	7.02	5.58E-09	8.25
13	49.00	0.00101	0.04949	0.0000	2.79E-08	7.55	5.15E-07	6.29	3.04E-08	7.52
14	50.00	0.00000	0.05000	0.0000	5.34E-06	5.27	2.29E-05	4.64	5.57E-06	5.25
15	51.00	0.00000	0.00000	9.901E-04	9.90E-04	3.00	9.90E-04	3.00	9.90E-04	3.00
16	55.00	0.00000	0.00000	4.762E-03	4.76E-03	2.32	4.76E-03	2.32	4.76E-03	2.32
17	60.00	0.00000	0.00000	9.091E-03	9.09E-03	2.04	9.09E-03	2.04	9.09E-03	2.04
18	**Spreadsheet Documentation**									
19	Applicable to all				For ammonia			For sodium cyanide		
20	Cell D2=$H$2/B2				Cell E7=$H$2/SQRT($B$7*$D$2)			Cell I7=$H$2/SQRT($B$7*$D$4)		
21	Cell D3=$H$2/B3				Cell F7=-LOG10(E7)			Cell J7=-LOG10(I7)		
22	Cell D4=$H$2/B4				Cell E8=$B$2*C8/B8			Cell I8 =$B$4*C8/B8		
23	Cell B7=($F$2*$B$5-$F$3*A7)/($B$5+A7)				Cell E14=SQRT(C14*$B$2)			Cell I14=SQRT(C14*$B$4)		
24	Cell B15=0 (entry)									
25	Cell C7=$F$3*A7/($B$5+A7)				For hydrazine					
26	Cell C15=0 (entry)				Cell G7=$H$2/SQRT($B$7*$D$3)					
27	Cell D7=0 (entry)				Cell H7=-LOG10(G7)					
28	Cell D15=(A15*$F$3-$B$5*$F$2)/($B$5+A15)				Cell G8=$B$3*C8/B8					
29	Cells E15, G15, I15=D15				Cell G14=SQRT(C14*$B$3)					

	A	B	C	D	E	F	G	H	I
1	**12-52 (a) and (c) titrations of 50.00 mL 0.1000 M acid with 0.1000 M NaOH**								
2	K_a anilinium chloride	2.51E-05	Initial conc. acid	0.1000	K_w	1.00E-14			
3	K_a for HOCl	3.00E-08	Initial conc. NaOH	0.1000					
4					(a)		(c)		
5	Vol. acid, mL	50.00			For anilinium		For hypochlorous		
6	**Vol. NaOH, mL**	c_{acid}	$c_{conj.base}$	c_{NaOH}	$[H_3O^+]$	pH	$[H_3O^+]$	pH	
7	0.00	0.10000	0.00000	0.0000	1.58E-03	2.80	5.48E-05	4.26	
8	5.00	0.08182	0.00909	0.0000	2.26E-04	3.65	2.70E-07	6.57	
9	15.00	0.05385	0.02308	0.0000	5.86E-05	4.23	7.00E-08	7.15	
10	25.00	0.03333	0.03333	0.0000	2.51E-05	4.60	3.00E-08	7.52	
11	40.00	0.01111	0.04444	0.0000	6.28E-06	5.20	7.50E-09	8.12	
12	45.00	0.00526	0.04737	0.0000	2.79E-06	5.55	3.33E-09	8.48	
13	49.00	0.00101	0.04949	0.0000	5.12E-07	6.29	6.12E-10	9.21	
14	50.00	0.00000	0.05000	0.0000	2.24E-09	8.65	7.75E-11	10.11	
15	51.00			0.0010	1.01E-11	11.00	1.01E-11	11.00	
16	55.00			0.0048	2.10E-12	11.68	2.10E-12	11.68	
17	60.00			0.0091	1.10E-12	11.96	1.10E-12	11.96	
18	**Spreadsheet Documentation**								
19	Applicable to both parts			Note that parts (a) and (c) use an identical method, and do not require the					
20	Cell B7=($D$2*$B$5-$D$3*A7)/($B$5+A7)			quadratic formula. Part (b) requires the quadratic formula and is shown on					
21	Cell C7=$D$3*A7/($B$5+A7)			a separate spreadsheet.					
22	Cell D7=0.000 (entry)								
23	Cell D15=(A15*$D$3-$B$5*$D$2)/($B$5+A15)			For hypochlorous acid					
24	Cell E15=$F$2/D15			Cell G7=SQRT($B$3*$D$2)					
25	For anilinium			Cell H7=-LOG10(G7)					
26	Cell E7=SQRT($B$2*$D$2)			Cell G8=$B$3*B8/C8					
27	Cell F7=-LOG10(E7)			Cell G14=$F$2/SQRT(C14*$F$2/$B$3)					
28	Cell E8=$B$2*B8/C8								
29	Cell E14=$F$2/SQRT(C14*$F$2/$B$2)								

	A	B	C	D	E	F	G	H
1	**12-52 (b) titration of 50.00 mL of 0.01000 M chloroacetic acid with 0.0100 M NaOH**							
2	K_a	1.36E-03	Initial conc. acid	0.0100	K_w	1.00E-14		
3			Intial conc. NaOH	0.0100				
4			Vol. acid, mL	50.00				
5	**Vol. NaOH, mL**	c_{acid}	$c_{conj.base}$	c_{NaOH}	$[H_3O^+]$	pH		
6	0.00	0.01000	0.00000	0.00000	3.07E-03	2.51		
7	5.00	0.00818	0.00091	0.00000	2.39E-03	2.62		
8	15.00	0.00538	0.00231	0.00000	1.44E-03	2.84		
9	25.00	0.00333	0.00333	0.00000	8.22E-04	3.09		
10	40.00	0.00111	0.00444	0.00000	2.50E-04	3.60		
11	45.00	0.00053	0.00474	0.00000	1.15E-04	3.94		
12	49.00	0.00010	0.00495	0.00000	2.17E-05	4.66		
13	50.00	0.00000	0.00500	0.00000	5.22E-08	7.28		
14	51.00			9.90E-05	1.01E-10	10.00		
15	55.00			4.76E-04	2.10E-11	10.68		
16	60.00			9.09E-04	1.10E-11	10.96		
17								
18	**Spreadsheet Documentation**			Note that the quadratic formula is required for this problem.				
19	Cell B6=($D$4*$D$2-A6*$D$3)/(A6+$D$4)							
20	Cell C6=$D$3*A6/(A6+$D$4)							
21	Cell D6=0.0000 (entry)							
22	Cell D14=((A14-$D$4)*$D$3)/(A14+$D$4)							
23	Cell E6=(-$B$2+SQRT($B$2^2-4*(-$D$2*$B$2)))/2							
24	Cell F6=-LOG10(E6)							
25	Cell E7=(-($B$2+C7)+SQRT(($B$2+C7)^2-4*(-$B$2*B7)))/2							
26	Cell E13=$F$2/(SQRT(C13*$F$2/$B$2))							
27	Cell E14=$F$2/D14							

	A	B	C	D	E	F	G	H
1	**12-52 (d) titration of 50.00 mL of 0.1000 M hydroxylamine with 0.1000 M HCl**							
2	K_a for hydroxyl ammonium	1.10E-06	K_b for hydroxylamine	9.09E-09	Initial conc. base	0.1000	K_w	1.00E-14
3					Initial conc. HCl	0.1000		
4	Vol. base, mL	50.00						
5	**Vol. HCl, mL**	c_{base}	$c_{conj.acid}$	c_{HCl}	$[H_3O^+]$	**pH**		
6	0.00	0.10000	0.00000	0.00000	3.32E-10	9.48		
7	5.00	0.08182	0.00909	0.00000	1.22E-07	6.91		
8	15.00	0.05385	0.02308	0.00000	4.71E-07	6.33		
9	25.00	0.03333	0.03333	0.00000	1.10E-06	5.96		
10	40.00	0.01111	0.04444	0.00000	4.40E-06	5.36		
11	45.00	0.00526	0.04737	0.00000	9.90E-06	5.00		
12	49.00	0.00101	0.04949	0.00000	5.39E-05	4.27		
13	50.00	0.00000	0.05000	0.00000	2.35E-04	3.63		
14	51.00			9.90E-04	9.90E-04	3.00		
15	55.00			4.76E-03	4.76E-03	2.32		
16	60.00			9.09E-03	9.09E-03	2.04		
17								
18	**Spreadsheet Documentation**		Note the method used in this problem is identical to that used in 12-51.					
19	Cell D2=$H$2/$B$2							
20	Cell B6=($F$2*$B$4-$F$3*A6)/($B$4+A6)							
21	Cell C6=$F$3*A6/($B$4+A6)							
22	Cell D6=0.0000(entry)							
23	Cell D14=(A14*$F$3-$B$4*$F$2)/(A14+$B$4)							
24	Cell E6=$H$2/SQRT($B$6*$D$2)							
25	Cell F6=-LOG10(E6)							
26	Cell E7=$B$2*C7/B7							
27	Cell E13==SQRT(C13*$B$2)							
28	Cell E14=D14							

12-53. Here we make use of Equations 12-16 and 12-17:

$$\alpha_0 = \frac{[H_3O^+]}{[H_3O^+]+K_a} \qquad \alpha_1 = \frac{K_a}{[H_3O^+]+K_a}$$

and obtain the results shown in the spreadsheet that follows.

	A	B	C	D	E	F	G	H	I	J	K	L	M
1	12-53			(a)	(b)	(c)	(d)	(e)					
2		K_a		acetic acid	picric acid	HOCl	hydroxylamine	piperdine					
3	Acetic	1.75E-05	pH	5.320	1.250	7.000	5.120	10.080					
4	Picric	4.30E-01	$[H_3O^+]$	4.786E-06	5.623E-02	1.000E-07	7.586E-06	8.318E-11					
5	HOCl	3.00E-08	α_0	0.215	0.116	0.769	0.873	0.917					
6	Hydroxylamine	1.10E-06	α_1	0.785	0.884	0.231	0.127	0.083					
7	Piperdine	7.50E-12	Note that in the table below numbers less than 0.001 are expressed in scientific notation.										
8			(a) acetic acid		(b) picric acid		(c) hypochlorous acid		(d) hydroxylammonium		(e) piperdinium		
9	pH	$[H_3O^+]$	α_0	α_1	α_0	α_1	α_0	α_1	α_0	α_1	α_0	α_1	
10	0.0	1.000	1.000	1.75E-05	0.699	0.301	1.000	3.00E-08	1.000	1.10E-06	1.000	7.50E-12	
11	0.5	0.316	1.000	5.53E-05	0.424	0.576	1.000	9.49E-08	1.000	3.48E-06	1.000	2.37E-11	
12	1.0	0.100	1.000	1.75E-04	0.189	0.811	1.000	3.00E-07	1.000	1.10E-05	1.000	7.50E-11	
13	1.5	0.032	0.999	5.53E-04	0.069	0.931	1.000	9.49E-07	1.000	3.48E-05	1.000	2.37E-10	
14	2.0	0.010	0.998	0.002	0.023	0.977	1.000	3.00E-06	1.000	1.10E-04	1.000	7.50E-10	
15	2.5	0.003	0.994	0.006	0.007	0.993	1.000	9.49E-06	1.000	3.48E-04	1.000	2.37E-09	
16	3.0	0.001	0.983	0.017	0.002	0.998	1.000	3.00E-05	0.999	0.001	1.000	7.50E-09	
17	3.5	3.16E-04	0.948	0.052	7.35E-04	0.999	1.000	9.49E-05	0.997	0.003	1.000	2.37E-08	
18	4.0	1.00E-04	0.851	0.149	2.33E-04	1.000	1.000	3.00E-04	0.989	0.011	1.000	7.50E-08	
19	4.5	3.16E-05	0.644	0.356	7.35E-05	1.000	0.999	9.48E-04	0.966	0.034	1.000	2.37E-07	
20	5.0	1.00E-05	0.364	0.636	2.33E-05	1.000	0.997	0.003	0.901	0.099	1.000	7.50E-07	
21	5.5	3.16E-06	0.153	0.847	7.35E-06	1.000	0.991	0.009	0.742	0.258	1.000	2.37E-06	
22	6.0	1.00E-06	0.054	0.946	2.33E-06	1.000	0.971	0.029	0.476	0.524	1.000	7.50E-06	
23	6.5	3.16E-07	0.018	0.982	7.35E-07	1.000	0.913	0.087	0.223	0.777	1.000	2.37E-05	
24	7.0	1.00E-07	0.006	0.994	2.33E-07	1.000	0.769	0.231	0.083	0.917	1.000	7.50E-05	
25	7.5	3.16E-08	0.002	0.998	7.35E-08	1.000	0.513	0.487	0.028	0.972	1.000	2.37E-04	
26	8.0	1.00E-08	5.71E-04	0.999	2.33E-08	1.000	0.250	0.750	0.009	0.991	0.999	7.49E-04	
27	8.5	3.16E-09	1.81E-04	1.000	7.35E-09	1.000	0.095	0.905	0.003	0.997	0.998	0.002	
28	9.0	1.00E-09	5.71E-05	1.000	2.33E-09	1.000	0.032	0.968	9.08E-04	0.999	0.993	0.007	
29	9.5	3.16E-10	1.81E-05	1.000	7.35E-10	1.000	0.010	0.990	2.87E-04	1.000	0.977	0.023	
30	10.0	1.00E-10	5.71E-06	1.000	2.33E-10	1.000	0.003	0.997	9.09E-05	1.000	0.930	0.070	
31	10.5	3.16E-11	1.81E-06	1.000	7.35E-11	1.000	0.001	0.999	2.87E-05	1.000	0.808	0.192	
32	11.0	1.00E-11	5.71E-07	1.000	2.33E-11	1.000	3.33E-04	1.000	9.09E-06	1.000	0.571	0.429	
33	11.5	3.16E-12	1.81E-07	1.000	7.35E-12	1.000	1.05E-04	1.000	2.87E-06	1.000	0.297	0.703	
34	12.0	1.00E-12	5.71E-08	1.000	2.33E-12	1.000	3.33E-05	1.000	9.09E-07	1.000	0.118	0.882	
35	12.5	3.16E-13	1.81E-08	1.000	7.35E-13	1.000	1.05E-05	1.000	2.87E-07	1.000	0.040	0.960	
36	13.0	1.00E-13	5.71E-09	1.000	2.33E-13	1.000	3.33E-06	1.000	9.09E-08	1.000	0.013	0.987	
37	13.5	3.16E-14	1.81E-09	1.000	7.35E-14	1.000	1.05E-06	1.000	2.87E-08	1.000	0.004	0.996	
38	14.0	1.00E-14	5.71E-10	1.000	2.33E-14	1.000	3.33E-07	1.000	9.09E-09	1.000	0.001	0.999	
39	**Spreadsheet Documentation**												
40	Cell D4=10^-D3		Cell E6=$E$3/(E4+$B$4)		Cell G5=G4/(G4+$B$6)		Cell B10=10^-A10		Cell G10=B10/(B10+$B$5)		Cell K10=B10/(B10+$B$7)		
41	Cell D5=D4/(D4+$B$3)		Cell F4=10^-F3		Cell G6=$B$6/(G4+$B$6)		Cell C10=B10/(B10+$B$3)		Cell H10=$B$5/(B10+$B$5)		Cell L10=$B$7/(B10+$B$7)		
42	Cell D6=$B$3/(D4+$B$3)		Cell F5=F4/(F4+$B$5)		Cell H4=10^-H3		Cell D10=$B$3/(B10+$B$3)		Cell I10 =B10/(B10+$B$6)				
43	Cell E4=10^-E3		Cell F6=$B$5/(F4+$B$5)		Cell H5=H4/(H4+$B$7)		Cell E10=B10/(B10+$B$4)		Cell J10=$B$6/(B10+$B$6)				
44	Cell E5=E4/(E4+$B$4)		Cell G4=10^-G3		Cell H6=$B$7/(H4+$B$7)		Cell F10=$B$4/(B10+$B$4)						

12-53.

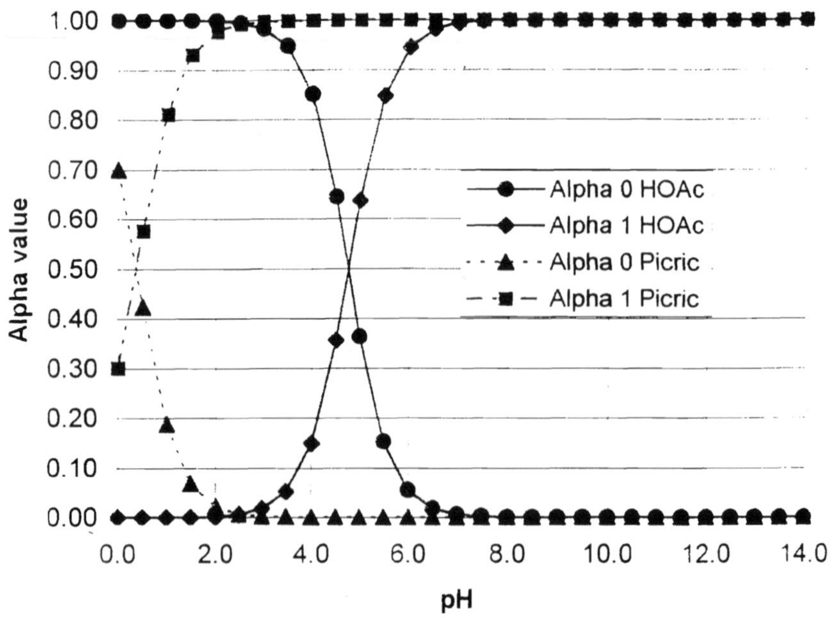

Alpha value
pH
Alpha 0 HOAc
Alpha 1 HOAc
Alpha 0 Picric
Alpha 1 Picric

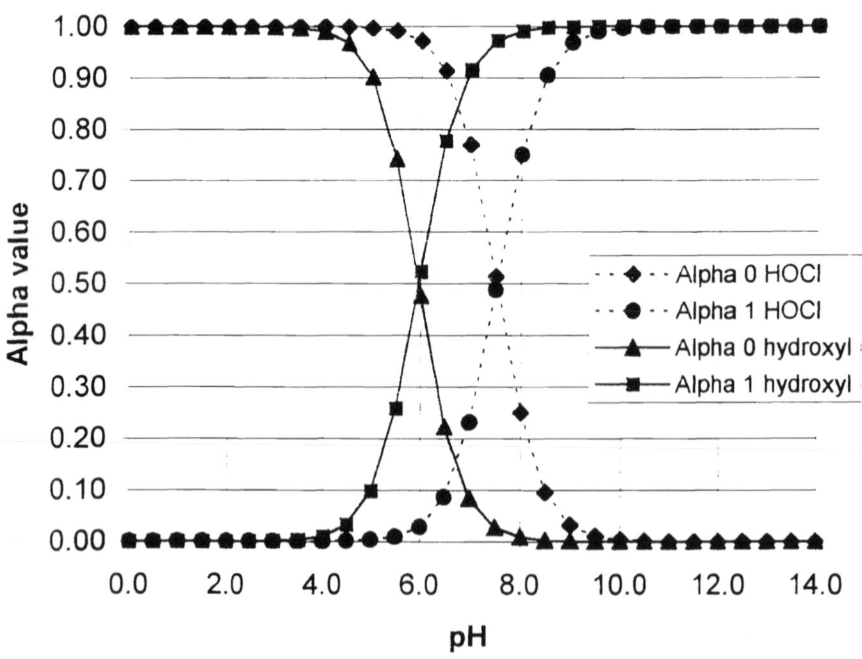

Alpha value
pH
Alpha 0 HOCl
Alpha 1 HOCl
Alpha 0 hydroxyl
Alpha 1 hydroxyl

12-53.

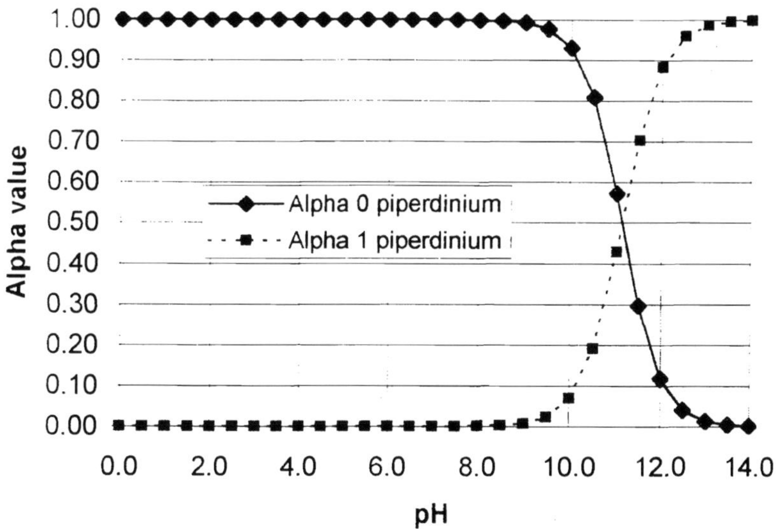
Alpha value
1.00
0.90
0.80
0.70
0.60
0.50
0.40
0.30
0.20
0.10
0.00
Alpha 0 piperdinium
Alpha 1 piperdinium
0.0
2.0
4.0
6.0
8.0
10.0
12.0
14.0
pH

12-54. $[H_3O^+] = 6.310 \times 10^{-4}$. Substituting into Equation 12-16 gives

$$\alpha_0 = \frac{6.310 \times 10^{-4}}{6.310 \times 10^{-4} + 1.80 \times 10^{-4}} = 0.778$$

$$\frac{\text{HCOOH}}{c_T} = \frac{[\text{HCOOH}]}{0.0850} = \alpha_0$$

$$[\text{HCOOH}] = 0.778 \times 0.0850 = \underline{\underline{6.61 \times 10^{-2}\ \text{M}}}$$

Chapter 13

13-1. Not only is NaHA a proton donor, it is also the conjugate base of the parent acid H_2A

$$HA^- + H_2O \rightleftharpoons H_3O^+ + A^{2-}$$
$$HA^- + H_2O \rightleftharpoons H_2A + OH^-$$

Solutions of acid salts are acidic or alkaline, depending upon which of these equilibrium predominates. In order to compute the pH of solutions of this type, it is necessary to take both equilibria into account.

13-3. The HPO_4^{2-} ion is such a weak acid ($K_a = 4.5 \times 10^{-13}$) that the change in pH in the vicinity of the third equivalence point is too small to be observable.

13-4. **(a)** $NH_4^+ + H_2O \rightleftharpoons NH_3 + H_3O^+ \qquad K_{NH_4^+} = 5.70 \times 10^{-10}$

$$OAc^- + H_2O \rightleftharpoons HOAc + OH^- \qquad K_{OAc^-} = \frac{K_w}{K_{HOAc}} = \frac{1.00 \times 10^{-14}}{5.70 \times 10^{-10}} = 5.71 \times 10^{-10}$$

Since the K's are essentially identical, the solution should be approximately neutral.

(c) Neither Na^+ nor NO_3^- reacts with H_2O. Solution will be neutral.

(e)
$$C_2O_4^{2-} + H_2O \rightleftharpoons HC_2O_4^- + OH^- \qquad K_b = \frac{1.00 \times 10^{-14}}{5.42 \times 19^{-5}} = 1.84 \times 10^{-10}$$

Solution will be basic.

(g) $H_2PO_4^- + H_2O \rightleftharpoons H_3O^+ + HPO_4^- \qquad K_2 = 6.32 \times 10^{-8}$

$$H_2PO_4^- + H_2O \rightleftharpoons OH^- + H_3PO_4 \qquad K_b = \frac{1.00 \times 10^{-14}}{7.11 \times 10^{-3}} = 1.41 \times 10^{-12}$$

Solution will be acidic because $K_2 > K_b$.

13-6. Let us assume that $c_{Na_2HAsO_4} = 0.033$ M at the equivalence point and that $K_2 = 1.1 \times 10^{-7}$ and $K_3 = 3.2 \times 10^{-12}$. Substituting into Equation 13-3 gives

$$[H_3O^+] = \sqrt{\frac{3.2 \times 10^{-12} \times 0.033 + 1 \times 10^{-14}}{1 + 0.033/1.1 \times 10^{-7}}} = 6.2 \times 10^{-10}$$

$pH = -\log(6.2 \times 10^{-10}) = 9.2$ Phenolphthalein would be suitable.

13-8. **(a)** To obtain the approximate equivalence point pH, we will employ Equation 13-4

$$[H_3O^+] = \sqrt{K_1K_2} = \sqrt{1.5 \times 10^{-4} \times 4.69 \times 10^{-11}}$$

$$= 8.39 \times 10^{-8} \quad \text{and} \quad pH = -\log(8.39 \times 10^{-8}) = 7.08$$

Bromothymol blue. (6.2 to 7.6) (See Table 12-1, page 268)

(c) $T^{2-} + H_2O \rightleftarrows HT^- + OH^-$

$$K_b = \frac{[OH^-][HT^-]}{[T^{2-}]} = \frac{1.00 \times 10^{-14}}{4.31 \times 10^{-5}} = 2.32 \times 10^{-10}$$

$$[OH^-] = \sqrt{0.0500 \times 1.00 \times 10^{-14}/4.31 \times 10^{-5}} = 3.41 \times 10^{-6}$$

$$[OH^-] = [HT^-] \quad \text{and} \quad T^{2-} = 0.050 - [OH^-] \approx 0.050$$

$$= \sqrt{0.050 \times 2.32 \times 10^{-10}} = 3.41 \times 10^{-6}$$

$pH = 14.00 - (-\log 3.41 \times 10^{-6}) = 8.5$ Cresol purple. (7.6 to 9.2)

(e) $^+H_3NCH_2CH_2NH_3^+ + H_2O \rightarrow H_2NCH_2CH_2NH_3^+ + H_3O^+ \qquad K_1 = 1.42 \times 10^{-7}$

$$\frac{[H_3O^+]^2}{0.0500} = 1.42 \times 10^{-7}$$

$$[H_3O^+ = \sqrt{1.42 \times 10^{-7} \times 0.0500} = 8.43 \times 10^{-5}$$

$pH = -\log(8.43 \times 10^{-5}) = 4.07$ Bromocresol green. (3.8 to 5.4)

(g) Proceeding as in part (c), we obtain pH = 9.94.

Phenolphthalein. (8.5 to 10.0)

13-9. (a) $H_3PO_4 + H_2O \rightleftarrows H_3O^+ + H_2PO_4^- \qquad K_1 = 7.11 \times 10^{-3}$

$$\frac{[H_3O^+][H_2PO_4^-]}{[H_3PO_4]} = \frac{[H_3O^+]^2}{0.04 - [H_3O^+]} = 7.11 \times 10^{-3}$$

$$[H_3O^+]^2 + 7.11 \times 10^{-3}[H_3O^+] - 0.0400 \times 7.11 \times 10^{-3} = 0$$

Solving by systematic approximation or by the quadratic formula gives

$[H_3O^+] = 1.37 \times 10^{-2}$ and pH = - log (1.37×10^{-2}) = 1.86

Proceeding in the same way, we obtain

(c) pH = 1.64

(e) pH = 4.21

13-10. In this problem, we will use Equation 13-3 or one of its simplifications.

(a)

$$[H_3O^+] = \sqrt{\frac{0.0400 \times 6.32 \times 10^{-8}}{1 + 0.0400/7.11 \times 10^{-3}}} \qquad K_2 c \gg K_w$$

$$= 1.95 \times 10^{-5} \quad \text{and} \quad \text{pH} = -\log(1.95 \times 10^{-5}) = 4.71$$

Proceeding as in part (a), we obtain

(c) pH = 4.28

(e) pH = 9.80 Here, $c_{HA}/K_1 >> 1$.

13-11. (a) $PO_4^{3-} + H_2O \rightleftarrows HPO_4^{2-} + OH^- \qquad K_b = \frac{K_w}{K_3} = \frac{1.00 \times 10^{-14}}{4.5 \times 10^{-13}} = 2.22 \times 10^{-2}$

$$\frac{[OH^-]^2}{0.0400 - [OH^-]} = 2.22 \times 10^{-2}$$

$$[OH^-]^2 + 2.22 \times 10^{-2}[OH^-] - 8.88 \times 10^{-4} = 0$$

Solving gives $[OH^-] = 2.07 \times 10^{-2}$

$$pH = 14.00 - (-\log 2.07 \times 10^{-2}) = \underline{\underline{12.32}}$$

Proceeding as in part (a), we obtain

(c) pH = $\underline{\underline{9.70}}$

(e) Proceeding as in part (a) gives pH = $\underline{\underline{12.58}}$

13-12. (a) $[H_3O^+][H_2AsO_4^-]/[H_3AsO_4] = 5.8 \times 10^{-3}$

Here we must use Equations 12-6 and 12-7 (page 277) because K_a is relatively large.

Thus,

$$[H_2AsO_4^-] = 0.0200 + [H_3O^+] - [OH^-]$$

$$[H_3AsO_4] = 0.0500 - [H_3O^+] + [OH^-]$$

Since $[OH^-]$ is negligible, we can write

$$[H_3O^+]\,(0.0200) + [H_3O^+])\,/\,(0.0500 - [H_3O^+]) = 5.8 \times 10^{-3}$$

$$[H_3O^+]^2 + 2.58 \times 10^{-2}\,[H_3O^+] - 2.90 \times 10^{-4} = 0$$

$[H_3O^+] = 8.46 \times 10^{-3}$ and pH = $\underline{\underline{2.07}}$

(b) $H_2AsO_4^- + H_2O \rightleftarrows H_3O^+ + HAsO_4^{2-}$

$$[H_3O^+][HAsO_4^{2-}]/[H_2AsO_4^-] = 1.11 \times 10^{-7}$$

$[HAsO_4^{2-}] \approx 0.0500$ and $[H_2AsO_4^-] \approx 0.0300$

$$[H_3O^+] = 1.1 \times 10^{-7} \times 0.0300/0.0500 = 6.6 \times 10^{-8}$$

pH = $-\log (6.6 \times 10^{-8}) = \underline{\underline{7.18}}$

(c) Proceeding as in part (b), we obtain pH = 10.63.

(d) $H_3PO_4 + HPO_4^{2-} \rightarrow 2H_2PO_4^-$

For each milliliter of solution, 0.0200 mmol Na_2HPO_4 reacts with 0.0200 mmol H_3PO_4 to give 0.0400 mmol NaH_2PO_4 and to leave 0.0200 mmol H_3PO_4. Thus we have a buffer that is 0.0200 M in H_3PO_4 and 0.0400 M in NaH_2PO_4. Since K_1 for H_3PO_4 is relatively large, we proceed as in part (a), and obtain pH = 2.55.

(e) Proceeding as in part (a), we obtain pH = 2.06.

13-14. (a) $HA + H_2O \rightleftarrows H_3O^+ + A^- \qquad K_a = 4.3 \times 10^{-1}$

Picric acid is a strong enough weak acid so that it contributes significantly to $[H_3O^+]$. Its contribution is given by $[A^-]$ and the total $[H_3O^+]$ is then given by

$$[H_3O^+] = c_{HCl} + [A^-] = 0.0100 + [A^-]$$

$$[A^-] = [H_3O^+] - 0.0100$$

Mass balance requires that $[HA] + [A^-] = 0.0200$, or

$$[HA] = 0.0200 - [A^-] = 0.0200 - [H_3O^+] + 0.0100$$

$$= 0.0300 - [H_3O^+]$$

$$\frac{[H_3O^+][A^-]}{[HA]} = 4.3 \times 10^{-1} = \frac{[H_3O^+]([H_3O^+] - 0.0100)}{0.0300 - [H_3O^+]}$$

$$[H_3O^+]^2 + 0.42\,[H_3O^+] - 1.29 \times 10^{-2} = 0$$

$[H_3O^+] = 2.87 \times 10^{-2}$ and pH = 1.54

(b) Proceeding in the same way, we obtain for benzoic acid, HBz

$$\frac{[H_3O^+]([H_3O^+] - 0.0100)}{0.0300 - [H_3O^+]} = 6.28 \times 10^{-5}$$

$$[H_3O^+]^2 - 9.94 \times 10^{-3}[H_3O^+] - 6.28 \times 10^{-5} \times 0.0300 = 0$$

$[H_3O^+] = 1.013 \times 10^{-2}$ and pH = $\underline{\underline{1.99}}$

Note that if we assumed that the HCl completely repressed the dissociation of HBz, the calculated pH would be 2.00.

(c) $CO_3^{2-} + H_2O \rightleftarrows HCO_3^- + OH^- \qquad K_b = \frac{K_w}{K_2} = \frac{1.00 \times 10^{-14}}{4.69 \times 10^{-11}} = 2.13 \times 10^{-4}$

Hydroxide ions are introduced from two sources; NaOH and Na_2CO_3. The concentration from the latter source is equal to $[HCO_3^-]$. Thus,

$$[OH^-] = c_{NaOH} + [HCO_3^-]$$

$$= 0.0100 + [HCO_3^-]$$

$$[HCO_3^-] = [OH^-] - 0.0100$$

Since the solution is 0.100 M in Na_2CO_3, we may write

$$[CO_3^{2-}] = 0.100 - [HCO_3^-] - [H_2CO_3] \approx 0.100 - [HCO_3^-]$$

Substituting the previous equation gives

$$[CO_3^{2-}] = 0.100 - ([OH^-] - 0.0100) = 0.110 - [OH^-]$$

$$\frac{[OH^-][HCO_3^-]}{[CO_3^{2-}]} = 2.13 \times 10^{-4} = \frac{[OH^-]([OH^-] - 0.0100)}{0.110 - [OH^-]}$$

$$2.343 \times 10^{-5} - 2.13 \times 10^{-4}[OH^-] = [OH^-]^2 - 0.0100[OH^-]$$

$$[OH^-]^2 - 9.79 \times 10^{-3}[OH^-] - 2.343 \times 10^{-5} = 0$$

$$[OH^-] = 1.178 \times 10^{-2}$$

pH = 14.00 − (-log 1.178×10^{-2}) = $\underline{\underline{12.07}}$

(d) $NH_3 + H_2O \rightleftarrows NH_4^+ + OH^- \qquad K_b = \frac{[NH_4^+][OH^-]}{[NH_3]} = \frac{1.00 \times 10^{-14}}{5.70 \times 10^{-10}} = 1.75 \times 10^{-5}$

Proceeding in the same way as part (c), we write

$$[OH^-] = 0.0100 + [NH_4^+] \quad \text{or} \quad [NH_4^+] = [OH^-] - 0.0100$$

$$[NH_4^+] + [NH_3] = 0.100 \quad \text{or} \quad [NH_3] = 0.100 - [NH_4^+]$$

$$\frac{[OH^-]([OH^-] - 0.0100)}{0.110 - [OH^-]} = 1.75 \times 10^{-5}$$

$$[OH^-]^2 - 9.983 \times 10^{-3}[OH^-] - 1.925 \times 10^{-6} = 0$$

$$[OH^-] = 1.0172 \times 10^{-2} \quad \text{and} \quad pH = 14.00 - \log(1.0172 \times 10^{-2}) = \underline{\underline{12.01}}$$

13-16. (a) Let us compare the ratio $[H_2SO_3]/[HSO_3^-]$ with that of $[SO_3^{2-}]/[HSO_3^-]$.

The larger will contain the predominant acid/base pair. The first is obtained by inverting the numerical value for K_1 and substituting $[H_3O^+] = 1.00 \times 10^{-6}$.

$$\frac{[H_2SO_3]}{[H_3O^+][HSO_3^-]} = \frac{1}{K_1} \quad \text{and} \quad \frac{[H_3O^+][SO_3^{2-}]}{[HSO_3^-]} = K_2$$

$$\frac{[H_2SO_3]}{[HSO_3^-]} = \frac{1.00 \times 10^{-6}}{1.23 \times 10^{-2}} = 8.1 \times 10^{-5} \qquad \frac{[SO_3^{2-}]}{[HSO_3^-]} = \frac{6.6 \times 10^{-8}}{1.00 \times 10^{-6}} = 0.066$$

Clearly the predominant pair is SO_3^{2-}/HSO_3^- and its acid/base ratio is

$1/0.066 \approx \underline{\underline{15.2}}$.

(b) Substituting $[H_3O^+] = 1.00 \times 10^{-6}$ into the expressions for K_1, K_2, and K_3 yields

$$\frac{[H_2Cit^-]}{[H_3Cit]} = 745 \qquad \frac{[HCit^{2-}]}{[H_2Cit^-]} = 17.3 \qquad \frac{[Cit^{3-}]}{[HCit^{2-}]} = 0.40$$

The large size of the first two ratios and the small size of the third indicate that $HCit^{2-}$ is a predominant species in this solution. To compare $[Cit^{3-}]$ and $[H_2Cit^-]$ we invert the second ratio. Then

$$[H_2Cit^-]/[HCit^{2-}] = 1/17.3 = 0.058$$

Thus, the predominant acid/base system involves $[Cit^{3-}]$ and $[HCit^{2-}]$ and their acid/base ratio is

$$[HCit^{2-}]/[Cit^{3-}] = 1/0.40 = \underline{\underline{2.5}}$$

(c) Proceeding as in part (a), we obtain $[HM^-]/[M^{2-}] = 0.498$

(d) Proceeding as in part (a), we obtain $[HT^-]/[T^{2-}] = 0.0232$

13-18. pH = 7.30 $\quad [H_3O^+] = \text{antilog}(-7.30) = 5.012\times10^{-8}$

$$[H_3O^+][HPO_4^{2-}]/[H_2PO_4^-] = 6.32\times10^{-8}$$

$$[HPO_4^{2-}]/[H_2PO_4^-] = 6.32\times10^{-8}/(5.012\times10^{-8}) = 1.261$$

$$HPO_4^{2-} + H_3PO_4 \rightarrow 2H_2PO_4^-$$

no. mmol H_3PO_4 present $= 400\times0.200 = 80.0$

no. mmol $H_2PO_4^{2-}$ in the buffer $= 2\times80.0 = 160.0$

no. mmol HPO_4^{2-} needed for the buffer $= 1.261\times160.0 = 201.8$

Thus we need 80.0 mL of Na_2HPO_4 to react with the H_3PO_4 and an additional 201.8 to provide the needed concentration of HPO_4^{2-} or 281.8 mmol.

$$\text{mass } Na_2HPO_4\cdot2H_2O = 281.8\text{ mmol}\times0.17799\text{ g/mmol} = \underline{\underline{50.2\text{ g}}}$$

13-20. no. mmol $NaH_2PO_4 = 50.0\times0.200 = 10.0$

(a) no. mmol H_3PO_4 formed = no. mmol HCl added $= 50.0\times0.120 = 6.00$

$c_{H_3PO_4} \;=\; 6.00/100 \;=\; 0.0600\text{ M}$

$c_{NaH_2PO_4} \;=\; (10.0 - 6.00)/100 \;=\; 0.0400$

Proceeding as in Problem 13-12(a), we obtain pH = 2.11.

(b) $c_{Na_2HPO_4} \;=\; 6.00/100 \;=\; 0.0600\text{ M}$

$c_{NaH_2PO_4} \;=\; (10.00 - 6.00)/100 \;=\; 0.0400$

Proceeding as in Problem 13-12(b), using K_2, gives pH = 7.38.

13-22. pH = 9.60 $[H_3O^+] = \text{antilog}(-9.60) = 2.512 \times 10^{-10}$

$[H_3O^+][CO_3^{2-}]/[HCO_3^-] \;=\; 4.69 \times 10^{-11}$

$[CO_3^{2-}]/[HCO_3^-] \;=\; 4.69 \times 10^{-11}/2.512 \times 10^{-10} \;=\; 0.1867$

Let V_{HCl} = mL 0.200 M HCl and $V_{Na_2CO_3}$ = mL 0.300 M Na_2CO_3

Since the solutions are dilute, the volumes will be additive

$V_{HCl} + V_{Na_2CO_3} \;=\; 1000\text{ mL}$

Assume

$[CO_3^{2-}] \;\approx\; c_{Na_2CO_3} \;=\; (V_{Na_2CO_3} \times 0.300 - V_{HCl} \times 0.200)/1000$

$[HCO_3^-] \;\approx\; c_{HCO_3^-} \;=\; V_{HCl} \times 0.200/1000$

Substituting these relationships into the ratio $[CO_3^{2-}]/[HCO_3^-]$ gives

$$\frac{0.300\,V_{Na_2CO_3} - 0.200\,V_{HCl}}{0.200\,V_{HCl}} \;=\; 0.1867$$

$0.300\,V_{Na_2CO_3} - 0.200\,V_{HCl} \;=\; 0.03734\,V_{HCl}$

$$0.300\,(1000 - V_{\text{HCl}}) = 0.23734\,V_{\text{HCl}}$$

$$V_{\text{HCl}} = 300/0.5373 = \underline{\underline{558\text{ mL}}}$$

$$V_{\text{Na}_2\text{CO}_3} = 1000 - 558 = \underline{\underline{442\text{ mL}}}$$

Thus mix 442 mL of 0.300 M Na_2CO_3 with (1000 - 442) = 558 mL of 0.200 M HCl.

13-28. (a) $K_{a1} = \dfrac{[H_3O^+][HCO_3^-]}{[H_2CO_3]} = 1.5 \times 10^{-4}$ $\quad K_{a2} = \dfrac{[H_3O^+][CO_3^{2-}]}{[HCO_3^-]} = 4.69 \times 10^{-11}$

0.00 mL

$$CO_3^{2-} + H_2O \rightleftarrows HCO_3^- + OH^-$$

$$K_{b1} = 1.00 \times 10^{-14}/4.69 \times 10^{-11} = 2.132 \times 10^{-4}$$

$$[OH^-][HCO_3^-]/[CO_3^{2-}] = K_{b1}$$

$$[OH^-] = [HCO_3^-] \quad \text{and} \quad [CO_3^{2-}] = 0.1000 - [OH^-] \approx 0.100$$

$$[OH^-]^2/0.1000 = K_{b1}$$

$$[OH^-] = \sqrt{K_{b1} \times 0.1000}$$

$$[H_3O^+] = K_w/[OH^-] = K_w/\sqrt{K_{b1} \times 0.1000}$$

$$\text{pH} = -\log[H_3O^+]$$

For this first point, the value of pH = 11.66 is found in cell G10 of the spreadsheet that follows.

12.50 mL

We now form a Na_2CO_3 / $NaHCO_3$ buffer in which

$$[HCO_3^-] \approx c_{\text{NaHCO}_3} = 12.50 \times 0.2000/62.50 = 4.000 \times 10^{-2}$$

$$[CO_3^{2-}] \approx c_{Na_2CO_3} = (50.0 \times 0.1000 - 12.50 \times 0.2000)/62.5$$

$$= 4.000 \times 10^{-2}$$

Substituting into the expressions for K_{a2} and rearranging give

$$[H_3O^+] = K_{a2} \times c_{NaHCO_3} / c_{Na_2CO_3}$$

pH = $\underline{\underline{10.33}}$ as shown in cell G11 of the spreadsheet.

20.00 and 24.00 mL

These data are treated in the same way. The results are found in the spreadsheet.

25.00 mL

Here we have a 0.06667 M solution of HCO_3^- and Equation 13-3 (page 305) applies. That is,

$$[H_3O^+] = \sqrt{\frac{K_{a2}\, c_{NaHCO_3} + K_w}{1 + c_{NaHCO_3}/K_{a1}}}$$

The pH = 7.08 as found in cell G14.

26.00 mL

We are now dealing with a new buffer mixture of H_2CO_3/$NaHCO_3$.

no. mmol HCl added = 26.00×0.200 = 5.200

no. mmol HCl consumed to form HCO_3^- = 50.00×0.1000 = $\underline{5.000}$

no. mmol H_2CO_3 formed = 0.200

no. mmol HCO_3^- remaining = 50.00×0.1000 - 0.2000 = 4.800

$$c_{H_2CO_3} = 0.2000/76.00 = 2.632 \times 10^{-3}$$

$$c_{NaHCO_3} = 4.800/76.00 = 6.316 \times 10^{-2}$$

Because K_{a1} is large we cannot approximate. Applying Equations 12-6 and 12-7 gives

$$[H_2CO_3] = c_{H_2CO_3} - [H_3O^+] + [OH^-]$$

$$[HCO_3^-] = c_{NaHCO_3} + [H_3O^+] - [OH^-]$$

Since the solution should be acidic, we can neglect $[OH^-]$ to obtain

$$K_{a1} = \frac{[H_3O^+]\left(c_{NaHCO_3} + [H_3O^+]\right)}{c_{H_2CO_3} - [H_3O^+]}$$

$$[H_3O^+]^2 + (c_{NaHCO_3} + K_{a1})[H_3O^+] - K_{a1}\, c_{H_2CO_3} = 0$$

The solution to this quadratic equation is shown in cell F15 of the spreadsheet, and pH = 5.21.

37.50, 45.00, and 49.00 mL

These data are treated in the same way. Answers in the spreadsheet that follows.

50.00 mL

Here we have a solution that is 0.0500 M in H_2CO_3 and the pH is calculated on the basis of K_{a1}.

$$K_{a1} = \frac{[H_3O^+][HCO_3^-]}{[H_2CO_3]}$$

$$[H_3O^+] = [HCO_3^-]$$

$$c_{H_2CO_3} = [H_2CO_3] + [HCO_3^-]$$

$$[H_2CO_3] = c_{H_2CO_3} - [HCO_3^-] = c_{H_2CO_3} - [H_3O^+]$$

$$K_{a1} = \frac{[H_3O^+]^2}{c_{H_2CO_3} - [H_3O^+]}$$

$$[H_3O^+]^2 + K_{a1}[H_3O^+] - K_{a1}c_{H_2CO_3} = 0$$

The result is shown in cell F19 of the spreadsheet and pH = 2.57.

51.00 mL

We now have excess HCl, but H_2CO_3 is dissociated enough to contribute to the $[H_3O^+]$. The excess HCl is

$$c_{HCl} = 1.00 \times 0.2000/101.0 = 1.98 \times 10^{-3}\text{ M}$$

$$[H_3O^+] = c_{HCl} + [HCO_3^-] \qquad [HCO_3^-] = [H_3O^+] - c_{HCl}$$

$$[H_2CO_3] = c_{H_2CO_3} - [H_3O^+]$$

$$K_{a1} = \frac{[H_3O^+]([H_3O^+] - c_{HCl})}{c_{H_2CO_3} - [H_3O^+]}$$

Solving for $[H_3O^+]$ gives

$$[H_3O^+]^2 + (K_{a1} - c_{HCl})[H_3O^+] - K_{a1}c_{H_2CO_3} = 0$$

The solution to this quadratic is shown in cell G20 of the spreadsheet and pH = 2.42.

60.00 mL

Now the acid is enough in excess that we can neglect the dissociation of H_2CO_3.

Here

$$c_{HCl} = \frac{\text{total mmol HCl added} - \text{mmol needed to titrate } Na_2CO_3}{\text{volume of solution}}$$

$$c_{HCl} = \frac{60.00 \times 0.2000 - 2 \times 0.1000 \times 50.00}{110.0} = 1.818 \times 10^{-2}$$

$$\text{pH} = -\log(1.818 \times 10^{-2}) = 1.74$$

(c) $HSO_4^- + H_2O \rightleftarrows H_3O^+ + SO_4^{2-}$ $[H_3O^+][SO_4^{2-}]/[HSO_4^-] = 1.02 \times 10^{-2}$

0.00 mL

Proceeding as in Feature 13-3 (page 314), we obtain

$$[H_3O^+] = 1.086 \times 10^{-1} \quad \text{and} \quad pH = \underline{\underline{0.96}}$$

as in cells F9 and G9 of the spreadsheet that follows.

12.50 mL

$$c_{NaHSO_4} = 12.50 \times 0.1000/62.50 = 0.04000$$

$$c_{H_2SO_4} = (50.00 \times 0.1000 - 12.50 \times 0.1000)/62.50 = 0.04000$$

$$[H_3O^+] = c_{H_2SO_4} + [SO_4^{2-}]$$

where $c_{H_2SO_4}$ represents the concentration of H_3O^+ from the complete dissociation of the H_2SO_4 to give HSO_4^-, and $[SO_4^{2-}]$ is equal to $[H_3O^+]$ from the partial dissociation of HSO_4^-.

Rearranging gives

$$[SO_4^{2-}] = [H_3O^+] - c_{H_2SO_4} \qquad (1)$$

Material balance requires that

$$[SO_4^{2-}] + [HSO_4^-] = c_{H_2SO_4} + c_{NaHSO_4} \qquad (2)$$

Substituting equation (1) into (2) gives upon rearranging

$$[HSO_4^-] = c_{H_2SO_4} + c_{NaHSO_4} + c_{H_2SO_4} - [H_3O^+]$$

$$= 2\, c_{H_2SO_4} + c_{NaHSO_4} - [H_3O^+] \qquad (3)$$

Substituting equations (1) and (3) into the acid dissociation constant expression gives

$$\frac{[H_3O^+]\left([H_3O^+] - c_{H_2SO_4}\right)}{\left(2\,c_{H_2SO_4} + c_{NaHSO_4} - [H_3O^+]\right)} = 1.02\times10^{-2} = 0.0102$$

$$[H_3O^+]^2 + (0.0102 - c_{H_2SO_4})[H_3O^+] - 0.0102\,(2\,c_{H_2SO_4} + c_{NaHSO_4}) = 0$$

The results are shown in cell F10 in the spreadsheet and pH = 1.28.

20.00 and 24.00 mL

These calculations are performed in the same way and lead to pH = 1.50 and pH = 1.63.

25.00 mL

$$c_{NaHSO_4} = 50.00\times0.1000/75.00 = 6.667\times10^{-2}$$

$$[H_3O^+] = [SO_4^{2-}] \quad \text{and} \quad [HSO_4^-] = 6.667\times10^{-2} - [H_3O^+]$$

$$[H_3O^+]^2/(6.67\times10^{-2} - [H_3O^+]) = 1.02\times10^{-2}$$

$$[H_3O^+]^2 + 1.02\times10^{-2}[H_3O^+] - 6.803\times10^{-4} = 0$$

The solution is in cell F13 and pH = 1.67.

26.00 mL

Proceeding as in part (a) for the concentration calculation of 26.00 mL, we find

$$c_{Na_2SO_4} = 2.632\times10^{-3} \quad \text{and} \quad c_{NaHSO_4} = 6.316\times10^{-2}$$

HSO_4^- is a sufficiently strong acid that we must employ Equations 12-6 and 12-7, page 277.

$$[SO_4^{2-}] = 2.632\times10^{-3} + [H_3O^+] - [OH^-]$$

$$[HSO_4^-] = 6.316\times10^{-2} - [H_3O^+] + [OH^-]$$

Since $[OH^-]$ is negligible, we can write

$$[H_3O^+](2.632\times10^{-3}+[H_3O^+])/(6.316\times10^{-2}-[H_3O^+]) = 1.02\times10^{-2}$$

$$[H_3O^+]^2+1.283\times10^{-2}[H_3O^+]-6.442\times10^{-4} = 0$$

The solution is in cell F14 and pH = 1.70.

37.50, 45.00, and 49.00 mL

These calculations are performed in the same way as that for 26.00 mL. The results are located in the spreadsheet.

50.00 mL

Here, $c_{Na_2SO_4} = 0.05000$

Proceeding as in Problem 13-11(b), we obtain pH = 7.35.

51.00 mL

Here,

$$c_{NaOH} = [OH^-] = (51.00\times0.2000-2\times50.00\times0.1000)/101.00$$

$$= 1.980\times10^{-3}$$

$$[H_3O^+] = K_w/[OH^-]$$

The results are in cell F19 and pH = 11.30.

60.00 mL

Proceeding in the same way, we obtain pH = 12.26.

	A	B	C	D	E	F	G	H	I
1	**13-28 (a) and (b) 50.00 mL 0.1000 M base with 0.2000 M HCl**								
2	K_{a1} for carbonic acid	1.50E-04	$K_{b2} = K_w/K_{a1}$	6.67E-11	Initial conc. base	0.1000	K_w	1.00E-14	
3	K_{a2} for carbonic acid	4.69E-11	$K_{b1} = K_w/K_{a2}$	2.13E-04	Initial conc. HCl	0.2000			
4	K_{a1} for ethylenediamine	1.42E-07	$K_{b2} = K_w/K_{a1}$	7.04E-08					
5	K_{a2} for ethylenediamine	1.18E-10	$K_{b1} = K_w/K_{a2}$	8.47E-05					
6	Vol. base, mL	50.00							
7	Let B symbolize the original base (carbonate or ethylenediamine), BH^+ the first protonated species and BH^{2+} the species with 2 protons.								
8						(a) for carbonate		(b) for ethylenediamine	
9	**Vol. HCl, mL**	c_B	c_{BH^+}	$c_{BH^{2+}}$	c_{HCl}	$[H_3O^+]$	**pH**	$[H_3O^+]$	**pH**
10	0.00	0.10000	0.00000		0.0000	2.17E-12	11.66	3.44E-12	11.46
11	12.50	0.04000	0.04000		0.0000	4.69E-11	10.33	1.18E-10	9.93
12	20.00	0.01429	0.05714		0.0000	1.88E-10	9.73	4.72E-10	9.33
13	24.00	0.00270	0.06486		0.0000	1.13E-09	8.95	2.83E-09	8.55
14	25.00	0.00000	0.06667		0.0000	8.39E-08	7.08	4.10E-09	8.39
15	26.00	0.00000	0.06316	0.002632	0.0000	6.23E-06	5.21	5.92E-09	8.23
16	37.50	0.00000	0.02857	0.028571	0.0000	1.48E-04	3.83	1.42E-07	6.85
17	45.00	0.00000	0.01053	0.042105	0.0000	5.62E-04	3.25	5.68E-07	6.25
18	49.00	0.00000	0.00202	0.048485	0.0000	1.82E-03	2.74	3.41E-06	5.47
19	50.00	0.00000	0.00000	0.050000	0.0000	2.66E-03	2.57	8.43E-05	4.07
20	51.00	0.00000	0.00000	0.049505	1.980E-03	3.79E-03	2.42	1.98E-03	2.70
21	60.00	0.00000	0.00000	0.045455	1.818E-02	1.82E-02	1.74	1.82E-02	1.74
22	**Spreadsheet Documentation**			For carbonate				Cell H19=SQRT($B$4*D19)	
23	Applicable to all			Cell F10=$H$2/SQRT($D$3*B10)				Cell H20=E20	
24	Cell D2=$H$2/$B$2			Cell F11=$B$3*C11/B11					
25	Cell D3=$H$2/$B$3			Cell F14=SQRT(($B$3*C14+$H$2)/(1+C14/$B$2))					
26	Cell D4=$H$2/$B$4			Cell F15=(-(C15+$B$2)+SQRT((C15+$B$2)^2-4*(-$B$2*D15)))/2					
27	Cell D5=$H$2/$B$5			Cell F19=(-$B$2+SQRT($B$2^2-4*(-$B$2*D19)))/2					
28	Cell B10=($F$2*$B$6-$F$3*A10)/($B$6+A10)			Cell F20=(-($B$2-E20)+SQRT(($B$2-E20)^2-4*(-$B$2*D20)))/2					
29	Cell B15 =0.0000 (entry)			Cell F21=E21					
30	Cell C10=$F$3*A10/($B$6+A10)			Cell G10=-LOG10(F10)					
31	Cell C15=($B$6*$F$2-(A15*$F$3-$B$6*$F$2))/(A15+$B$6)			For ethylene diamine					
32	Cell C20 = 0.0000 (entry)			Cell H10=$H$2/SQRT($D$5*B10)					
33	Cell D15=(A15*$F$3-$B$6*$F$2)/(A15+$B$6)			Cell H11=$B$5*C11/B11					
34	Cell D19=($A$19*$F$3-$B$6*$F$2)/(A19+$B$6)			Cell H14=SQRT(($B$5*C14+$H$2)/(1+C14/$B$4))					
35	Cell E20=($F$3*A20-2*$F$2*$B$6)/($B$6+A20)			Cell H15=$B$4*D15/C15					

13-28.

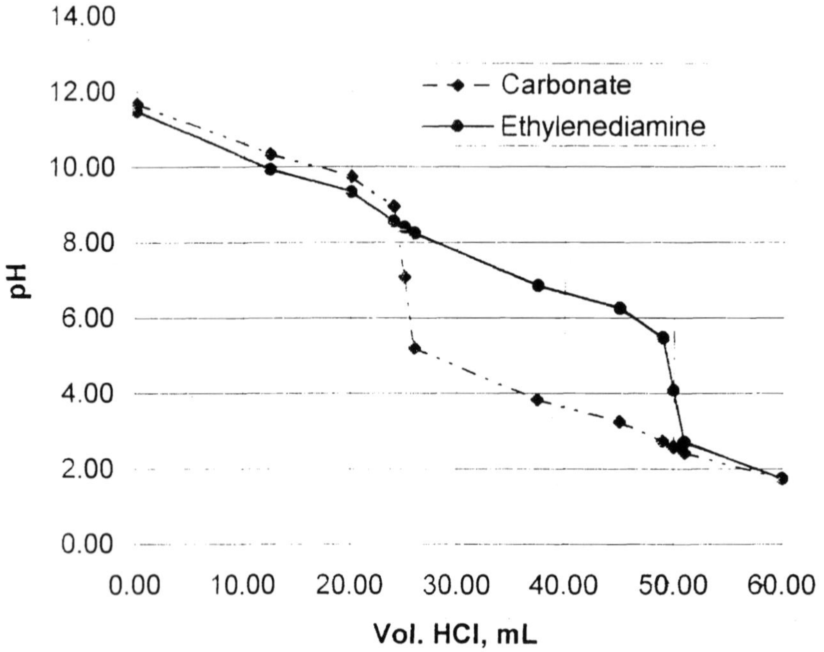
14.00
12.00
10.00
8.00
6.00
4.00
2.00
0.00
pH
Carbonate
Ethylenediamine
0.00
10.00
20.00
30.00
40.00
50.00
60.00
Vol. HCl, mL

13-28. (c), (d)

A	B	C	D	E	F	G	H	I	J
13-28 (c) and (d)	**50.00 mL 0.1000 M acid with 0.2000 M NaOH**								
K_{a1} sulfuric	Strong			Initial conc. acid	0.1000	K_w	1.00E-14		
K_{a2} sulfuric	1.02E-02	K_{b1} sulfuric	9.80E-13	Initial conc. NaOH	0.2000				
K_{a1} sulfurous	1.23E-02								
K_{a2} sulfurous	6.60E-08	K_{b1} sulfurous	1.52E-07						
Vol. acid, mL	50.00								
					(c) For sulfuric			(d) For sulfurous	
Vol. NaOH	c_{H2A}	c_{HA^-}	$c_{A^{2-}}$	**[OH⁻]**	**[H₃O⁺]**	**pH**	**[OH⁻]**	**[H₃O⁺]**	**pH**
0.00	0.10000				0.10859	0.96		0.02946	1.53
12.50	0.04000	0.04000			0.05293	1.28		0.00814	2.09
20.00	0.01429	0.05714			0.03168	1.50		0.00244	2.61
24.00	0.00270	0.06486			0.02328	1.63		4.28E-04	3.37
25.00	0.00000	0.06667			0.02147	1.67		2.62E-05	4.58
26.00	0.00000	0.06316	0.002632		0.01976	1.70		1.58E-06	5.80
37.50	0.00000	0.02857	0.028571		0.00645	2.19		6.60E-08	7.18
45.00	0.00000	0.01053	0.042105		0.00198	2.70		1.65E-08	7.78
49.00	0.00000	0.00202	0.048485		0.00035	3.46		2.75E-09	8.56
50.00	0.00000	0.00000	0.050000	2.21E-07	4.52E-08	7.35	8.70E-05	1.15E-10	9.94
51.00	0.00000	0.00000		1.98E-03	5.05E-12	11.30	1.98E-03	5.05E-12	11.30
60.00	0.00000	0.00000		1.82E-02	5.50E-13	12.26	1.82E-02	5.50E-13	12.26

Spreadsheet Documentation

applicable to all
Cell B9=$B$6*$F$2/($B$6+A9)
Cell B10=($B$6*$F$2-$F$3*A10)/($B$6+A10)
Cell B14=0.0000(entry)
Cell C10=$F$3*A10/($B$6+A10)
Cell C18=0.0000(entry)
Cell C14=($B$6*$F$2-($F$3*A14-$F$2*$B$6))/($B$6+A14)
Cell D14=($F$3*A14-$F$2*$B$6)/($B$6+A14)
Cell D18=$F$2*A18/($B$6+A18)
Cell E19, H19=(A19*$F$3-2*$B$6*$F$2)/($B$6+A19)
For sulfuric
Cell F9=(-($B$3-B9)+SQRT(($B$3-B9)^2-4*(-$B$3*2*B9)))/2
Cell G9=-LOG10(F9)
Cell F10=(-($B$3-B10)+SQRT(($B$3-B10)^2-4*(-$B$3*(2*B10+C10))))/2
Cell F13=(-$B$3+SQRT($B$3^2-4*(-$B$3*C13)))/2
Cell F14=(-(D14+$B$3)+SQRT((D14+$B$3)^2-4*(-$B$3*C14)))/2
Cell E18=SQRT(D18*$D$3)
Cell F18=$H$2/E18
For sulfurous
Cell I9=(-$B$4+SQRT($B$4^2-4*(-$B$4*B9)))/2
Cell J9=-LOG10(I9)
Cell I10=(-(C10+$B$4)+SQRT((C10+$B$4)^2-4*(-$B$4*B10)))/2
Cell I13=SQRT(($B$5*C13+$H$2)/(1+C13/$B$4))
Cell I14=$B$5*C14/D14
Cell H18=SQRT(D18*$D$5)
Cell I18=$H$2/H18

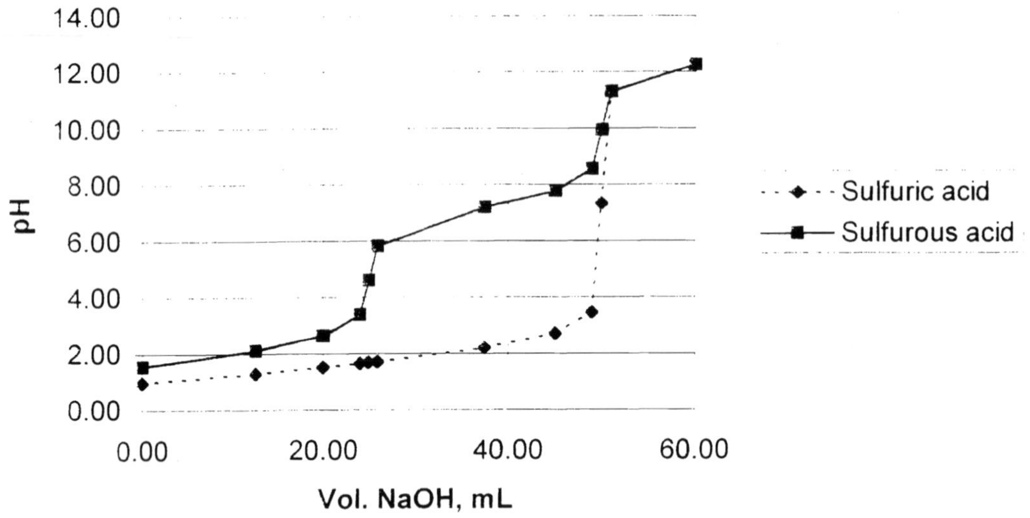
14.00
12.00
10.00
8.00
6.00
4.00
2.00
0.00
pH
0.00
20.00
40.00
60.00
Vol. NaOH, mL
Sulfuric acid
Sulfurous acid

13-29. $H_2NNH_3^+ + H_2O \rightleftarrows H_3O^+ + H_2NNH_2 \qquad K_a = 1.05 \times 10^{-8}$

$$H_2NNH_2 + H_2O \rightleftarrows H_2NNH_3^+ + OH^- \qquad K_b = \frac{1.00 \times 10^{-14}}{1.05 \times 10^{-8}} = 9.52 \times 10^{-7}$$

0.00 mL

$$c_{NaOH} = 0.1000 \approx [OH^-]$$

$$[H_3O^+] = K_w/[OH^-]$$

This calculation assumes that H_2NNH_2 contributes essentially no OH^- to the solution. The result is in cell F6 in the spreadsheet that follows and pH = 13.00.

10.00 mL

$$c_{NaOH} = (50.00 \times 0.1000 - 10.00 \times 0.200)/60.0 = 5.00 \times 10^{-2}$$

Assume $[OH^-] \approx c_{NaOH} = 5.00 \times 10^{-2}$

The results for $[H_3O^+]$ are in cell F7 of the spreadsheet, pH = 12.70

20.00 and 24.00 mL

Proceeding in the same way, we obtain the results in cells F8 and F9 and pH = 12.15 and 11.43.

25.00 mL

$$c_{H_2NNH_2} = 50.00 \times 0.008000/75.00 = 5.333 \times 10^{-2} - [OH^-]$$

$$c_{NaOH} = 0.000$$

$$[OH^-][H_2NNH_3^+]/[H_2NNH_2] = 9.52 \times 10^{-7}$$

$$[OH^-] = [H_2NNH_3^+]$$

$$[H_2NNH_2] = 5.333 \times 10^{-2} - [OH^-] \approx 5.333 \times 10^{-2}$$

$[OH^-] = \sqrt{K_b \times c_B}$

$[H_3O^+] = K_w / [OH^-]$

The results are in cell F10 and pH = 10.36.

26.00 mL

no. mmol $HClO_4$ added $= 26.00 \times 0.200 = 5.2000$

initial no. mmol NaOH $= 50.00 \times 0.100 = \underline{5.0000}$

no. mmol $H_2NNH_3^+$ formed $= 0.2000$

initial no. mmol $H_2NNH_2 = 50.00 \times 0.0800 = 4.000$

no. mmol H_2NNH_2 present $= 4.000 - 0.2000 = 3.800$

$c_{H_2NH_3^+} = 0.2000/76.00 = 2.632 \times 10^{-3} \approx [H_2NNH_3^+]$

$c_{H_2NNH_2} = 3.800/76.00 = 5.000 \times 10^{-2} \approx [H_2NNH_2]$

$[H_3O^+] = K_a \times c_{H_2NNH_3^+} / c_{H_2NNH_2}$

The results are in cell F11 and pH = 9.26.

35.00 and 44.00 mL

Proceeding in the same way, we obtain the results in cells F12 and F13 with pH = 7.98 and 6.70 respectively.

45.00 mL

$[H_3O^+] = \sqrt{K_a c_{BH^+}}$ (cell F14) pH = 4.68

46.00 and 50.00 mL

$[H_3O^+] \approx c_{HClO_4}$ (cells F15 and F16)

This is obtained from the excess $HClO_4$ added and assumes that dissociation of BH^+ is negligible.

	A	B	C	D	E	F	G	H	I
1	**13-29 50.00 mL of 0.1000 M NaOH and 0.0800 M hydrazine with 0.2000 M $HClO_4$**								
2	K_a for hydrazinium	1.05E-08	$K_b = K_w/K_a$	9.52E-07	Initial conc. NaOH	0.1000			
3	Initial volume soln.	50.00			Initial conc. Hydrazine	0.0800			
4	Conc. $HClO_4$	0.2000				K_w	1.00E-14		
5	**Vol. $HClO_4$, mL**	c_B	c_{BH^+}	c_{NaOH}	**$[OH^-]$**	**$[H_3O^+]$**	**pH**		
6	0.00			0.10000	0.1000	1.00E-13	13.00		
7	10.00			0.05000	0.0500	2.00E-13	12.70		
8	20.00			0.01429	0.0143	7.00E-13	12.15		
9	24.00			0.00270	0.0027	3.70E-12	11.43		
10	25.00	0.05333		0.00000	2.25E-04	4.44E-11	10.35		
11	26.00	0.05000	0.00263	0.00000		5.53E-10	9.26		
12	35.00	0.02353	0.02353	0.00000		1.05E-08	7.98		
13	44.00	0.00213	0.04043	0.00000		2.00E-07	6.70		
14	45.00	0.00000	0.04211	0.00000		2.10E-05	4.68		
15	46.00		0.04167	0.00000		2.08E-03	2.68		
16	50.00		0.04000	0.00000		1.00E-02	2.00		
17									
18	**Spreadsheet Documentation**								
19	Cell D2=$G$4/$B$2								
20	Cell D6=($F$2*$B$3-$B$4*A6)/($B$3+A6)								
21	Cell D11=0.0000(entry)								
22	Cell B10=($B$3*$F$3)/($B$3+A10)								
23	Cell B11=($B$3*$F$3-($B$4*A11-$B$3*$F$2))/($B$3+A11)								
24	Cell B15=0.0000(entry)								
25	Cell C11=($B$4*A11-$B$3*$F$2)/($B$3+A11)								
26	Cell C14=$F$3*$B$3/($B$3+A14)								
27	Cell E6=D6								
28	Cell E10=SQRT($D$2*B10)								
29	Cell F6=$G$4/E6								
30	Cell F11=$B$2*C11/B11								
31	Cell F14=SQRT(C14*$B$2)								
32	Cell F15=($B$4*A15-($B$3*$F$3+$B$3*$F$2))/($B$3+A15)								
33	Cell G6=-LOG10(F6)								
34									

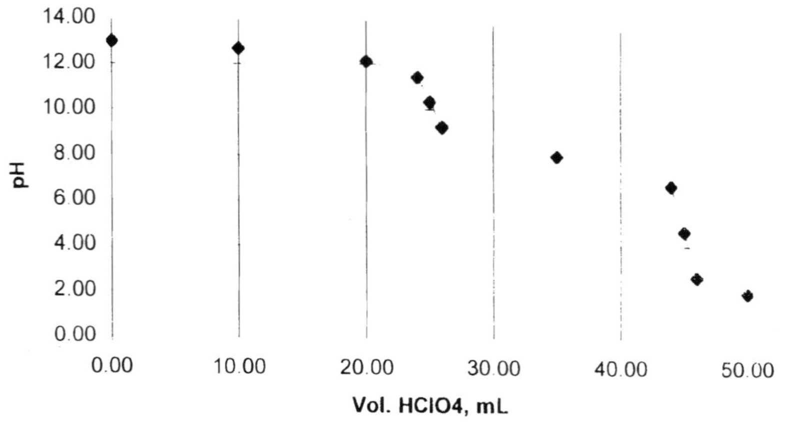

	A	B	C	D	E	F	G
1	**13-30 50.00 mL of 0.1000 M $HClO_4$ and 0.0800 M formic acid with 0.2000 M KOH**						
2	K_a for formic acid	1.80E-04	$K_b = K_w/K_a$	5.56E-11	Initial conc. $HClO_4$	0.1000	
3	Initial volume soln.	50.00			Initial conc. HCOOH	0.0800	
4	Conc. KOH	0.2000				K_w	1.00E-14
5	**Vol. KOH, mL**	c_{HA}	c_{A^-}	c_{HClO4}	**[OH^-]**	**[H_3O^+]**	**pH**
6	0.00	0.08000		0.10000		1.00E-01	1.00
7	10.00	0.06667		0.05000		5.00E-02	1.30
8	20.00	0.05714		0.01429		1.50E-02	1.82
9	24.00	0.05405		0.00270		4.70E-03	2.33
10	25.00	0.05333		0.00000		3.01E-03	2.52
11	26.00	0.05000	0.00263	0.00000		1.91E-03	2.72
12	35.00	0.02353	0.02353	0.00000		1.77E-04	3.75
13	44.00	0.00213	0.04043	0.00000		9.43E-06	5.03
14	45.00		0.04211	0.00000	1.53E-06	6.54E-09	8.18
15	46.00			0.00000	2.08E-03	4.80E-12	11.32
16	50.00			0.00000	1.00E-02	1.00E-12	12.00
17							
18	**Spreadsheet Documentation**						
19	Cell D2=$G$4/$B$2						
20	Cell B6=$B$3*$F$3/($B$3+A6)						
21	Cell B11=($B$3*$F$3-(A11*$B$4-$B$3*$F$2))/($B$3+A11)						
22	Cell C11=(A11*$B$4-$B$3*$F$2)/($B$3+A11)						
23	Cell D6=($F$2*$B$3-$B$4*A6)/($B$3+A6)						
24	Cell D11=0.00000(entry)						
25	Cell E14=SQRT(C14*$D$2)						
26	Cell E15=(A15*$B$4-($B$3*$F$3+$B$3*$F$2))/(A15+$B$3)						
27	Cell F6=D6						
28	Cell F8= (-($B$2-D8)+SQRT(($B$2-D8)^2-4*(-$B$2*(B8+D8))))/2						
29	Cell F10=(-$B$2+SQRT($B$2^2-4*(-$B$2*B10)))/2						
30	Cell F11=(-($B$2+C11)+SQRT(($B$2+C11)^2-4*(-$B$2*B11)))/2						
31	Cell F14=$G$4/E14						
32	Cell G6=-LOG10(F6)						

13-31. **(a)** $2\,H_2AsO_4^- \rightleftharpoons H_3AsO_4 + HAsO_4^{2-}$

$$K_1 = \frac{[H_3O^+][H_2AsO_4^-]}{[H_3AsO_4]} = 5.8\times10^{-3} \qquad (1)$$

$$K_2 = \frac{H_3O^+][HAsO_4^{2-}]}{[H_2AsO_4^-]} = 1.1\times10^{-7} \qquad (2)$$

$$K_3 = \frac{[H_3O^+][AsO_4^{3-}]}{[HAsO_4^{2-}]} = 3.2\times10^{-12} \qquad (3)$$

Dividing equation (2) by equation (1) leads to

$$\frac{K_2}{K_1} = \frac{[H_3AsO_4][HAsO_4^{2-}]}{[H_2AsO_4^-]^2} = \underline{\underline{1.9\times10^{-5}}}$$

which is the desired equilibrium-constant expression.

13-32. $HOAc + H_2O \rightleftharpoons H_3O^+ + OAc^- \qquad K_{HOAc} = 1.75\times10^{-5}$

$NH_4^+ + H_2O \rightleftharpoons H_3O^+ + NH_3 \qquad K_{NH_4^+} = 5.70\times10^{-10}$

Subtracting the first reaction from the second and rearranging give

$NH_4^+ + OAc^- \rightleftharpoons NH_3 + HOAc \qquad K = K_{NH_4^+}/K_{HOAc}$

$$\frac{[NH_3][HOAc]}{[NH_4^+][OAc^-]} = \frac{5.70\times10^{-10}}{1.75\times10^{-5}} = 3.26\times10^{-5}$$

	A	B	C	D	E	G	H	I	K	L	M	N
1	**13-33(a), (e) and (f)**		**Alpha values for various diprotic acids**									
2		K_{a1}	K_{a2}			Equations 13-12, 13-13 and 13-14 are used to obtain the alpha values in the						
3	Phthalic	1.12E-03	3.91E-06			spreadsheet below.						
4	Phosphorous	3.00E-02	1.62E-07			Note: The chart that follows shows only phthalic acid.						
5	Oxalic	5.60E-02	5.42E-05			The documentation for this spreadsheet is shown with the chart.						
6				(a)			(e)			(f)		
7				For phthalic acid			For phosphorous acid			For oxalic acid		
8	**pH**	**$[H_3O^+]$**	α_0	α_1	α_2	α_0	α_1	α_2	α_0	α_1	α_2	
9	0.0	1.0000	0.999	1.12E-03	4.37E-09	0.971	0.029	4.72E-09	0.947	0.053	2.87E-06	
10	0.5	0.3162	0.996	3.53E-03	4.36E-08	0.913	0.087	4.44E-08	0.850	0.150	2.58E-05	
11	1.0	0.1000	0.989	0.011	4.33E-07	0.769	0.231	3.74E-07	0.641	0.359	1.95E-04	
12	1.5	0.0316	0.966	0.034	4.23E-06	0.513	0.487	2.49E-06	0.361	0.638	1.09E-03	
13	2.0	0.0100	0.899	0.101	3.94E-05	0.250	0.750	1.21E-05	0.151	0.845	4.58E-03	
14	2.5	0.0032	0.738	0.261	3.23E-04	0.095	0.905	4.63E-05	0.053	0.931	0.016	
15	3.0	0.0010	0.471	0.527	2.06E-03	0.032	0.968	1.57E-04	0.017	0.933	0.051	
16	3.5	3.16E-04	0.218	0.772	9.55E-03	0.010	0.989	5.07E-04	4.80E-03	0.850	0.146	
17	4.0	1.00E-04	0.079	0.886	0.035	3.32E-03	0.995	0.002	1.16E-03	0.648	0.351	
18	4.5	3.16E-05	0.025	0.868	0.107	1.05E-03	0.994	0.005	2.08E-04	0.368	0.631	
19	5.0	1.00E-05	6.38E-03	0.714	0.279	3.28E-04	0.984	0.016	2.78E-05	0.156	0.844	
20	5.5	3.16E-06	1.26E-03	0.447	0.552	1.00E-04	0.951	0.049	3.11E-06	0.055	0.945	
21	6.0	1.00E-06	1.82E-04	0.204	0.796	2.87E-05	0.861	0.139	3.23E-07	0.018	0.982	
22	6.5	3.16E-07	2.11E-05	0.075	0.925	6.97E-06	0.661	0.339	3.28E-08	5.80E-03	0.994	
23	7.0	1.00E-07	2.23E-06	0.025	0.975	1.27E-06	0.382	0.618	3.29E-09	1.84E-03	0.998	
24	7.5	3.16E-08	2.27E-07	8.02E-03	0.992	1.72E-07	0.163	0.837	3.29E-10	5.83E-04	0.999	
25	8.0	1.00E-08	2.28E-08	2.55E-03	0.997	1.94E-08	0.058	0.942	3.29E-11	1.84E-04	1.000	
26	8.5	3.16E-09	2.28E-09	8.08E-04	0.999	2.02E-09	0.019	0.981	3.29E-12	5.83E-05	1.000	
27	9.0	1.00E-09	2.28E-10	2.56E-04	1.000	2.04E-10	6.13E-03	0.994	3.29E-13	1.84E-05	1.000	
28	9.5	3.16E-10	2.28E-11	8.09E-05	1.000	2.05E-11	1.95E-03	0.998	3.29E-14	5.83E-06	1.000	
29	10.0	1.00E-10	2.28E-12	2.56E-05	1.000	2.06E-12	6.17E-04	0.999	3.29E-15	1.85E-06	1.000	
30	10.5	3.16E-11	2.28E-13	8.09E-06	1.000	2.06E-13	1.95E-04	1.000	3.29E-16	5.83E-07	1.000	
31	11.0	1.00E-11	2.28E-14	2.56E-06	1.000	2.06E-14	6.17E-05	1.000	3.29E-17	1.85E-07	1.000	
32	11.5	3.16E-12	2.28E-15	8.09E-07	1.000	2.06E-15	1.95E-05	1.000	3.29E-18	5.83E-08	1.000	
33	12.0	1.00E-12	2.28E-16	2.56E-07	1.000	2.06E-16	6.17E-06	1.000	3.29E-19	1.85E-08	1.000	
34	12.5	3.16E-13	2.28E-17	8.09E-08	1.000	2.06E-17	1.95E-06	1.000	3.29E-20	5.83E-09	1.000	
35	13.0	1.00E-13	2.28E-18	2.56E-08	1.000	2.06E-18	6.17E-07	1.000	3.29E-21	1.85E-09	1.000	
36	13.5	3.16E-14	2.28E-19	8.09E-09	1.000	2.06E-19	1.95E-07	1.000	3.29E-22	5.83E-10	1.000	
37	14.0	1.00E-14	2.28E-20	2.56E-09	1.000	2.06E-20	6.17E-08	1.000	3.29E-23	1.85E-10	1.000	

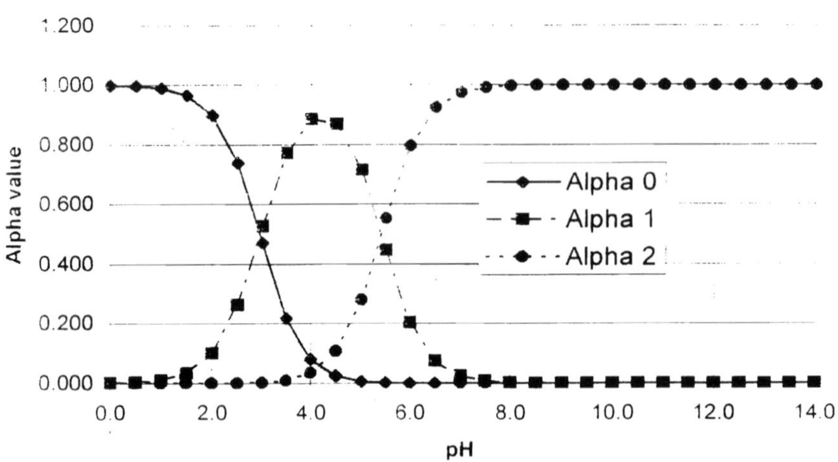

Documentation for diprotic acid alpha value spreadsheet

Cell B9=10^-A9

Cell C9=(B9^2)/(B9^2+$B$3*B9+$B$3*$C$3)

Cell D9=($B$3*B9)/(B9^2+$B$3*B9+$B$3*$C$3)

Cell E9=($B$3*$C$3)/(B9^2+$B$3*B9+$B$3*$C$3)

Cell G9=(B9^2)/(B9^2+$B$4*B9+$B$4*$C$4)

Cell H9=($B$4*B9)/(B9^2+$B$4*B9+$B$4*$C$4)

Cell I9=($B$4*$C$4)/(B9^2+$B$4*B9+$B$4*$C$4)

Cell K9=(B9^2)/(B9^2+$B$5*B9+$B$5*$C$5)

Cell L9=($B$5*B9)/(B9^2+$B$5*B9+$B$5*$C$5)

Cell M9=($B$5*$C$5)/(B9^2+$B$5*B9+$B$5*$C$5)

	A	B	C	D	E	F	H	I	J	K	L	M
1	**13-33(b), and (c)**	**Alpha values for various triprotic acids**										
2		K_{a1}	K_{a2}	K_{a3}			Feature 13-5 shows the calculation method for the alpha					
3	Phosphoric	7.11E-03	6.32E-08	4.50E-13			values in the spreadsheet.					
4	Citric	7.45E-04	1.73E-05	4.02E-07			Note: The chart that follows shows only phosphoric acid.					
5							The documentation for this spreadsheet is shown with the chart.					
6					(b)				(c)			
7				For phosphoric acid					For citric acid			
8	**pH**	**$[H_3O^+]$**	α_0	α_1	α_2	α_3	α_0	α_1	α_2	α_3		
9	0.0	1.000	0.993	7.06E-03	4.46E-10	2.01E-22	0.999	7.44E-04	1.29E-08	5.18E-15		
10	0.5	0.316	0.978	0.022	4.39E-09	6.25E-21	0.998	2.35E-03	1.29E-07	1.63E-13		
11	1.0	0.100	0.934	0.066	4.20E-08	1.89E-19	0.993	7.39E-03	1.28E-06	5.14E-12		
12	1.5	0.032	0.816	0.184	3.67E-07	5.22E-18	0.977	0.023	1.26E-05	1.60E-10		
13	2.0	1.00E-02	0.584	0.416	2.63E-06	1.18E-16	0.931	0.069	1.20E-04	4.82E-09		
14	2.5	3.16E-03	0.308	0.692	1.38E-05	1.97E-15	0.808	0.190	1.04E-03	1.32E-07		
15	3.0	1.00E-03	0.123	0.877	5.54E-05	2.49E-14	0.569	0.424	7.33E-03	2.95E-06		
16	3.5	3.16E-04	0.043	0.957	1.91E-04	2.72E-13	0.287	0.676	0.037	4.70E-05		
17	4.0	1.00E-04	0.014	0.986	6.23E-04	2.80E-12	0.103	0.765	0.132	5.32E-04		
18	4.5	3.16E-05	4.42E-03	0.994	1.99E-03	2.83E-11	0.027	0.626	0.343	4.36E-03		
19	5.0	1.00E-05	1.40E-03	0.992	6.27E-03	2.82E-10	4.77E-03	0.355	0.615	0.025		
20	5.5	3.16E-06	4.36E-04	0.980	0.020	2.79E-09	5.92E-04	0.139	0.763	0.097		
21	6.0	1.00E-06	1.32E-04	0.940	0.059	2.67E-08	5.31E-05	0.040	0.685	0.275		
22	6.5	3.16E-07	3.71E-05	0.833	0.167	2.37E-07	3.39E-06	7.98E-03	0.437	0.555		
23	7.0	1.00E-07	8.62E-06	0.613	0.387	1.74E-06	1.54E-07	1.15E-03	0.199	0.800		
24	7.5	3.16E-08	1.48E-06	0.333	0.666	9.48E-06	5.66E-09	1.33E-04	0.073	0.927		
25	8.0	1.00E-08	1.92E-07	0.137	0.863	3.89E-05	1.88E-10	1.40E-05	0.024	0.976		
26	8.5	3.16E-09	2.12E-08	0.048	0.952	1.36E-04	6.06E-12	1.43E-06	7.80E-03	0.992		
27	9.0	1.00E-09	2.19E-09	0.016	0.984	4.43E-04	1.93E-13	1.43E-07	2.48E-03	0.998		
28	9.5	3.16E-10	2.21E-10	4.97E-03	0.994	1.41E-03	6.10E-15	1.44E-08	7.86E-04	0.999		
29	10.0	1.00E-10	2.21E-11	1.57E-03	0.994	4.47E-03	1.93E-16	1.44E-09	2.49E-04	1.000		
30	10.5	3.16E-11	2.19E-12	4.93E-04	0.985	0.014	6.10E-18	1.44E-10	7.87E-05	1.000		
31	11.0	1.00E-11	2.13E-13	1.51E-04	0.957	0.043	1.93E-19	1.44E-11	2.49E-05	1.000		
32	11.5	3.16E-12	1.95E-14	4.38E-05	0.875	0.125	6.10E-21	1.44E-12	7.87E-06	1.000		
33	12.0	1.00E-12	1.53E-15	1.09E-05	0.690	0.310	1.93E-22	1.44E-13	2.49E-06	1.000		
34	12.5	3.16E-13	9.18E-17	2.07E-06	0.413	0.587	6.10E-24	1.44E-14	7.87E-07	1.000		
35	13.0	1.00E-13	4.05E-18	2.88E-07	0.182	0.818	1.93E-25	1.44E-15	2.49E-07	1.000		
36	13.5	3.16E-14	1.46E-19	3.29E-08	0.066	0.934	6.10E-27	1.44E-16	7.87E-08	1.000		
37	14.0	1.00E-14	4.84E-21	3.44E-09	0.022	0.978	1.93E-28	1.44E-17	2.49E-08	1.000		

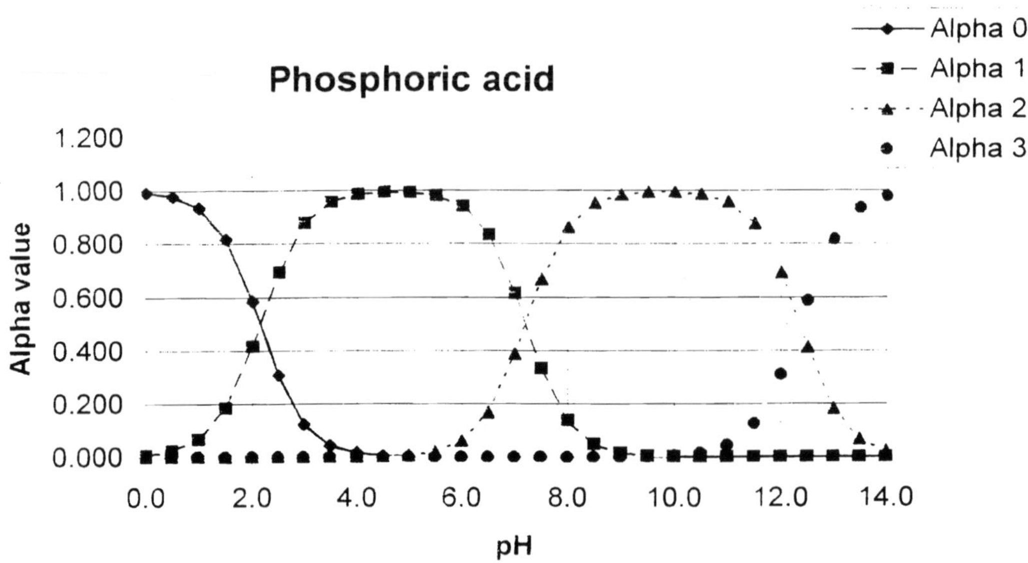

Cell B9=10^-A9
Cell C9=(B9^3)/(B9^3+$B$3*B9^2+$B$3*$C$3*B9+$B$3*$C$3*$D$3)
Cell D9=($B$3*B9^2)/(B9^3+$B$3*B9^2+$B$3*$C$3*B9+$B$3*$C$3*$D$3)
Cell E9=($B$3*$C$3*B9)/(B9^3+$B$3*B9^2+$B$3*$C$3*B9+$B$3*$C$3*$D$3)
Cell F9=($B$3*$C$3*$D$3)/(B9^3+$B$3*B9^2+$B$3*$C$3*B9+$B$3*$C$3*$D$3)
Cell H9=(B9^3)/(B9^3+$B$4*B9^2+$B$4*$C$4*B9+$B$4*$C$4*$D$4)
Cell I9=($B$4*B9^2)/(B9^3+$B$4*B9^2+$B$4*$C$4*B9+$B$4*$C$4*$D$4)
Cell J9=($B$4*$C$4*B9)/(B9^3+$B$4*B9^2+$B$4*$C$4*B9+$B$4*$C$4*$D$4)
Cell K9=($B$4*$C$4*$D$4)/(B9^3+$B$4*B9^2+$B$4*$C$4*B9+$B$4*$C$4*$D$4)

	A	B	C	D	E	F	G	H
1	**13-33 (d)**	**Alpha values for arsenic acid**						
2		K_{a1}	K_{a2}	K_{a3}	See Feature 13-5 for the equations used			
3	Arsenic	5.80E-03	1.10E-07	3.20E-12				
4	**pH**	**$[H_3O^+]$**	α_0	α_1	α_2	α_3		
5	0.0	1.000	0.994	5.77E-03	6.34E-10	2.03E-21		
6	0.5	0.316	0.982	0.018	6.27E-09	6.34E-20		
7	1.0	0.100	0.945	0.055	6.03E-08	1.93E-18		
8	1.5	0.032	0.845	0.155	5.39E-07	5.46E-17		
9	2.0	0.010	0.633	0.367	4.04E-06	1.29E-15		
10	2.5	0.003	0.353	0.647	2.25E-05	2.28E-14		
11	3.0	1.00E-03	0.147	0.853	9.38E-05	3.00E-13		
12	3.5	3.16E-04	0.052	0.948	3.30E-04	3.34E-12		
13	4.0	1.00E-04	0.017	0.982	1.08E-03	3.46E-11		
14	4.5	3.16E-05	5.40E-03	0.991	3.45E-03	3.49E-10		
15	5.0	1.00E-05	1.70E-03	0.987	0.011	3.48E-09		
16	5.5	3.16E-06	5.27E-04	0.966	0.034	3.40E-08		
17	6.0	1.00E-06	1.55E-04	0.901	0.099	3.17E-07		
18	6.5	3.16E-07	4.04E-05	0.742	0.258	2.61E-06		
19	7.0	1.00E-07	8.21E-06	0.476	0.524	1.68E-05		
20	7.5	3.16E-08	1.22E-06	0.223	0.777	7.86E-05		
21	8.0	1.00E-08	1.44E-07	0.083	0.916	2.93E-04		
22	8.5	3.16E-09	1.52E-08	0.028	0.971	9.83E-04		
23	9.0	1.00E-09	1.55E-09	8.98E-03	0.988	3.16E-03		
24	9.5	3.16E-10	1.55E-10	2.84E-03	0.987	9.99E-03		
25	10.0	1.00E-10	1.52E-11	8.80E-04	0.968	0.031		
26	10.5	3.16E-11	1.42E-12	2.61E-04	0.908	0.092		
27	11.0	1.00E-11	1.19E-13	6.89E-05	0.758	0.242		
28	11.5	3.16E-12	7.79E-15	1.43E-05	0.497	0.503		
29	12.0	1.00E-12	3.73E-16	2.16E-06	0.238	0.762		
30	12.5	3.16E-13	1.41E-17	2.59E-07	0.090	0.910		
31	13.0	1.00E-13	4.75E-19	2.75E-08	0.030	0.970		
32	13.5	3.16E-14	1.53E-20	2.81E-09	9.79E-03	0.990		
33	14.0	1.00E-14	4.88E-22	2.83E-10	3.12E-03	0.997		
34								
35	The documentation is shown on the chart that follows.							

13-33.

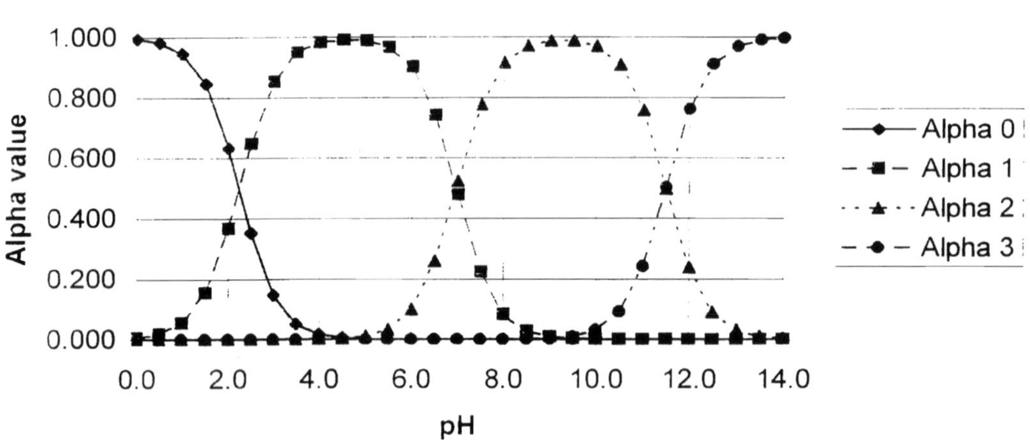

Cell B5=10^-A5
Cell C5=(B5^3)/(B5^3+$B$3*B5^2+$B$3*$C$3*B5+$B$3*$C$3*$D$3)
Cell D5=($B$3*B5^2)/(B5^3+$B$3*B5^2+$B$3*$C$3*B5+$B$3*$C$3*$D$3)
Cell E5=($B$3*$C$3*B5)/(B5^3+$B$3*B5^2+$B$3*$C$3*B5+$B$3*$C$3*$D$3)
Cell F5=($B$3*$C$3*$D$3)/(B5^3+$B$3*B5^2+$B$3*$C$3*B5+$B$3*$C$3*$D$3)

Chapter 14

14-1. Carbon dioxide is not strongly bonded by water molecules, and thus is readily volatilized from aqueous media. Gaseous HCl molecules, on the other hand, are fully dissociated into H_3O^+ and Cl^- when dissolved in water; neither of these species is volatile.

14-3. Primary standard Na_2CO_3 can be obtained by heating primary standard grade $NaHCO_3$ for about an hour at 270 to 300°C. The reaction is

$$2NaHCO_3(s) \rightarrow Na_2CO_3(s) + H_2O(g) + CO_2(g)$$

14-5. For, let us say, a 40-mL titration of $KH(IO_3)_2$

$$40 \text{ mL NaOH} \times 0.010 \frac{\text{mmol}}{\text{mL}} \times \frac{1 \text{ mmol KH(IO}_3)_2}{\text{mmol NaOH}} \times \frac{0.390 \text{ g KH(IO}_3)_2}{\text{mmol}} = 0.16 \text{ g KH(IO}_3)_2$$

and for titration of HBz

$$40 \text{ mL NaOH} \times 0.010 \frac{\text{mmol}}{\text{mL}} \times \frac{1 \text{ mmol HBz}}{\text{mmol NaOH}} \times \frac{0.122 \text{ g HBz}}{\text{mmol}} = 0.049 \text{ g HBz}$$

The $KH(IO_3)_2$ is preferable because the relative weighing error would be less with a 0.16-g sample than with a 0.049-g sample. A second reason for preferring $KH(IO_3)_2$ is because it is a strong acid and HBz is not. The titration error would therefore be less.

14-8. (a)
$$2.00 \text{ L KOH} \times \frac{0.15 \text{ mol KOH}}{\text{L}} \times \frac{56.1 \text{ g KOH}}{\text{mol}} = \underline{\underline{17 \text{ g KOH}}}$$

Dissolve 17 g KOH and dilute to 2.0 L.

(b)
$$2.00 \text{ L} \times 0.015 \frac{\text{mol Ba(OH)}_2 \cdot 8H_2O}{\text{L}} \times \frac{315 \text{ g Ba(OH)}_2 \cdot 8H_2O}{\text{mol}} = \underline{\underline{9.46 \text{ g}}}$$

Dissolve 9.5 g $Ba(OH)_2 \cdot 8H_2O$ in H_2O and dilute to 2.0 L.

(c)
$$2.00 \text{ L HCl} \times 0.200 \frac{\text{mol HCl}}{\text{L HCl}} \times 36.46 \frac{\text{g HCl}}{\text{mol}} = 14.58 \text{ g HCl needed}$$

$$14.58 \text{ g HCl} \times \frac{1 \text{ mL reagent}}{1.058 \text{ g reagent}} \times \frac{100 \text{ g reagent}}{11.50 \text{ g HCl}} = \underline{\underline{119.8 \text{ mL reagent}}}$$

Dilute about 120 mL of the reagent to 2.00 L.

14-10. For the first data set,

$$c_1 = 0.7987 \text{ g KHP} \times \frac{1 \text{ mmol KHP}}{0.20422 \text{ g KHP}} \times \frac{1 \text{ mmol NaOH}}{\text{mmol KHP}} \times \frac{1}{38.29 \text{ mL NaOH}} =$$

$$0.10214 \frac{\text{mmol NaOH}}{\text{mL}}$$

The data shown below for c_i were obtained in the same way.

Sample	c_i, M	c_i^2
1	0.10214	0.01043258
2	0.10250	0.01050625
3	0.10305	0.01061930
4	0.10281	0.01056990
	$\sum c_i = 0.41050$	$\sum c_i^2 = 0.04212803$

(a) $$\bar{c}_i = \frac{0.41050}{4} = \underline{\underline{0.1026 \text{ M}}}$$

(b) $$s = \sqrt{\frac{0.04212803 - (0.41050)^2/4}{4-1}} = \sqrt{\frac{0.000000466}{3}} = \underline{\underline{0.00039}}$$

$$\text{CV} = \frac{0.00039}{0.1026} \times 100\% = \underline{\underline{0.38\%}}$$

(c) Spread = 0.10305 - 0.10214 = 0.00091

$$Q = \frac{|0.10214 - 0.10250|}{0.00091} = 0.396$$

$Q_{crit} = 0.829$ at 95% confidence level

$Q_{crit} = 0.926$ at 99% confidence level

Hence, the first value is not an outlier.

14-12. **(a)** With phenolphthalein, the CO_3^{2-} consumes but 1 mmol H_3O^+ per mmol of CO_3^{2-}.

Thus the effective amount of base is lowered by 11.2 mmol, and

$$c_{base} = \frac{1000 \text{ mL NaOH} \times 0.1500 \frac{\text{mmol}}{\text{mL}} - 11.2 \text{ mmol } CO_2 \times \frac{1 \text{ mmol NaOH}}{\text{mmol } CO_2}}{1000}$$

$$= \underline{\underline{0.1388 \text{ M}}}$$

(b) When bromocresol green is the indicator,

$$CO_3^{2-} + 2H_3O^+ \rightarrow H_2CO_3 + 2H_2O$$

and the effective concentration of the base is unchanged. Thus,

$$c_{base} = \underline{\underline{0.1500 \text{ M}}}$$

14-14. **(a)**

$$0.6010 \text{ g AgCl} \times \frac{1 \text{ mmol AgCl}}{0.14332 \text{ g AgCl}} \times \frac{1 \text{ mmol HCl}}{\text{mmol AgCl}} \times \frac{1}{50.00 \text{ mL HCl}} = \underline{\underline{0.08387 \frac{\text{mmol HCl}}{\text{mL}}}}$$

(b)

$$25.00 \text{ mL Ba(OH)}_2 \times 0.04010 \frac{\text{mmol Ba(OH)}_2}{\text{mL Ba(OH)}_2} \times \frac{2 \text{ mmol HCl}}{\text{mmol Ba(OH)}_2} \times \frac{1}{19.92 \text{ mL HCl}} = \underline{\underline{0.1007 \frac{\text{mmol HCl}}{\text{mL HCl}}}}$$

(c)

$$0.2694 \text{ g } Na_2CO_3 \times \frac{1 \text{ mmol } Na_2CO_3}{0.10599 \text{ g } Na_2CO_3} \times \frac{2 \text{ mmol HCl}}{\text{mmol } Na_2CO_3} \times \frac{1}{38.77 \text{ mL HCl}} = \underline{\underline{0.1311 \frac{\text{mmol HCl}}{\text{mL HCl}}}}$$

14-16. **(a)** For 35 mL

$$35\ \text{mL HClO}_4 \times 0.150 \frac{\text{mmol HClO}_4}{\text{mL HClO}_4} \times \frac{1\ \text{mmol Na}_2\text{CO}_3}{2\ \text{mmol HClO}_4} \times \frac{0.10599\ \text{g Na}_2\text{CO}_3}{\text{mmol Na}_2\text{CO}_3} = 0.28\ \text{g Na}_2\text{CO}_3$$

Substituting 45 mL gives 0.36 g Na_2CO_3.

Thus, the range is 0.28 to 0.36 g Na_2CO_3.

Proceeding in the same way, we obtain

(c) 0.85 to 1.1 g of HBz

(e) 0.17 to 0.22 g TRIS

14-17. For TRIS and the 20.00 mL volume, the relative standard deviation in the molarity is the same as the relative standard deviation in the weight or

$$\text{RSD} = \left(\frac{0.0001}{0.048}\right) = 0.0021$$

The standard deviation in molarity is the relative standard deviation in the weight multiplied by the molarity (0.020 M). That is,

$$\text{SD} = \left(\frac{0.0001\ \text{g}}{0.048\ \text{g}}\right) \times 0.020\ \text{M} = 0.00004\ \text{M}$$

These same equations are used for the other volumes. The same procedure is used for Na_2CO_3 and $Na_2B_4O_7{\cdot}H_2O$. The results are shown in the tables below.

mL HCl	SD TRIS	RSD TRIS	SD Na_2CO_3	RSD Na_2CO_3
20.00	0.00004	0.0021	0.00009	0.0047
30.00	0.00003	0.0014	0.00006	0.0031
40.00	0.00002	0.0010	0.00005	0.0024
50.00	0.00002	0.0008	0.00004	0.0019

mL HCl	SD $Na_2B_4O_7 \cdot H_2O$	RSD $Na_2B_4O_7 \cdot H_2O$
20.00	0.00003	0.0013
30.00	0.00002	0.0009
40.00	0.00001	0.0007
50.00	0.00001	0.0005

14-19. $$21.48\text{ mL NaOH} \times 0.03776\frac{\text{mmol NaOH}}{\text{mL NaOH}} \times \frac{1\text{ mmol }H_2T}{2\text{ mmol NaOH}} \times \frac{0.15009\text{ g }H_2T}{\text{mmol }H_2T} \times \frac{100\text{ mL}}{50\text{ mL}} =$$

$$\underline{\underline{0.1217\text{ g }H_2T/100\text{ mL}}}$$

14-21. For each part, we may write

$$31.64\text{ mL HCl} \times 0.1081\frac{\text{mmol HCl}}{\text{mL HCl}} \times \frac{1}{0.7439\text{ g sample}} = 4.5978\frac{\text{mmol HCl}}{\text{g sample}}$$

(a) $$0.45978\frac{\text{mmol HCl}}{\text{g sample}} \times \frac{1\text{ mmol }Na_2B_4O_7}{2\text{ mmol HCl}} \times 0.20122\frac{\text{g }Na_2B_4O_7}{\text{mmol }Na_2B_4O_7} \times 100\% =$$

$$\underline{\underline{46.25\%\ Na_2B_4O_7}}$$

Proceeding in the same way

(b) $$0.45978 \times \frac{1}{2} \times 0.38137 \times 100 = 87.67\%\ Na_2B_4O_7 \cdot 10H_2O$$

(c) $$0.45978 \times 0.06962 \times 100 = 32.01\%\ B_2O_3$$

(d) $$0.45978\frac{\text{mmol HCl}}{\text{g sample}} \times \frac{2\text{ mmol B}}{\text{mmol HCl}} \times 0.010811\frac{\text{g B}}{\text{mmol B}} \times 100\% = 9.94\%\text{ B}$$

14-23. no. mmol NaOH = no. mmol HCHO + 2 × no. mmol H_2SO_4

no. mmol HCHO = 50.0 × 0.0996 - 2 × 23.3 × 0.05250 = 2.5335

$$\frac{2.5335 \text{ mmol HCHO} \times 0.030026 \text{ g HCHO/mmol HCHO}}{0.3124 \text{ g sample}} \times 100\% = \underline{\underline{24.4\% \text{ HCHO}}}$$

14-25. Letting RS_4 represent the compound,

$$1 \text{ mmol } RS_4 \equiv 4 \text{ mmol } SO_2 \equiv 4 \text{ mmol } H_2SO_4 \equiv 8 \text{ mmol NaOH}$$

$$\frac{22.13 \text{ mL NaOH} \times 0.03736 \frac{\text{mmol NaOH}}{\text{mL}} \times \frac{1 \text{ mmol } RS_4}{8 \text{ mmol NaOH}} \times 0.29654 \frac{\text{g } RS_4}{\text{mmol}}}{0.4329 \text{ g sample}} \times 100\% = \underline{\underline{7.079\% \text{ } RS_4}}$$

14-27. no. mmol HCl = no. mmol NaOH + 2 × no. mmol CO_3^{2-}

no. mmol CO_3^{2-} = (50.00 × 0.1140 - 24.21 × 0.09802) / 2 = 1.6635

molar mass CO_3^{2-} = 1 × 12.01115 + 3 × 15.9994 = 60.01

$$\text{molar mass of the salt} = \frac{0.1401 \text{ g salt}}{1.6635 \text{ mmol salt}} \times \frac{10^3 \text{ mmol}}{\text{mol}} = 84.22 \text{ g/mol}$$

molar mass of the cation of the salt = 84.22 - 60.01 = 24.21

$\underline{\underline{MgCO_3}}$ with a molar mass of 84.31 seems a likely candidate.

14-29. no. mmol $Ba(OH)_2$ = no. mmol CO_2 + no. mmol HCl/2

no. mmol CO_2 = 50.0 × 0.0116 - 23.6 × 0.0108/2 = 0.4526

$$\frac{0.4526 \text{ mmol } CO_2 \times 0.04401 \text{ g } CO_2\text{/mmol}}{3.00 \text{ L sample}} \times \frac{1 \text{ L } CO_2}{1.98 \text{ g } CO_2} \times 10^6 \text{ ppm} = \underline{\underline{3.35 \times 10^3 \text{ ppm}}}$$

14-31. 1 mmol P ≡ 1 mmol $(NH_4)_3PO_4 \cdot 12MoO_3$ ≡ 26 mmol NaOH

no. mmol NaOH = no. mmol HCl + 26 × no. mmol P

no. mmol P = (50.00 × 0.2000 - 14.17 × 0.1741)/26 = 0.28973

$$\frac{0.28973 \text{ mmol P} \times 0.030974 \text{ g P/mmol}}{0.1417 \text{ g sample}} \times 100\% = \underline{\underline{6.333\% \text{ P}}}$$

14-32. $C_6H_4(COOCH_3)_2 + 2OH^- \rightarrow 2CH_3OH + 2C_6H_4(COO^-)_2$

no. mmol NaOH = 2 × no. mmol analyte + no. mmol HCl

$$\text{no. mmol analyte} = (\text{no. mmol NaOH} - \text{no. mmol HCl})/2$$

$$= (50.00 \times 0.1031 - 24.27 \times 0.1644)/2 = 0.58251$$

$$\frac{0.58251 \text{ mmol analyte} \times 0.19419 \text{ g analyte/mmol}}{0.8160 \text{ g sample}} \times 100\% = \underline{\underline{13.86\% \text{ analyte}}}$$

14-33. 1 mmol RN_4 ≡ 4 mmol NH_3 ≡ 4 mmol HCl

$$\frac{(26.13 \times 0.01477) \text{ mmol HCl} \times \frac{1 \text{ mmol RN}_4}{4 \text{ mmol HCl}} \times 0.28537 \frac{\text{g RN}_4}{\text{mmol RN}_4}}{0.1247 \text{ g sample}} \times 100\% = 22.08\% \text{ RN}_4$$

14-34.

	A	B	C	D	E	F	G	H
1	**14-34 Tablets of quanidine**							
2	Molar mass quanidine	59.07						
3	Sample wt., g	7.5						
4	No. of tablets	4						
5	Volume of HCl, mL	100.00						
6	Conc. HCl, M	0.1750						
7	Vol. NaOH, mL	11.37						
8	Conc. NaoH, M	0.1080						
9	Conversion, kg/lb	0.454	In Cell B11, the no. mmoles of HCl consumed is calculated from					
10			the no. of mmoles of HCl added - no. of mmoles of NaOH needed.					
11	mmol HCl consumed	16.27204						
12	mg CH_5N_3/tablet	80.0991	In B12, the mg CH_5N_3/tablet is calculated from the mmol HCl consumed					
13			divided by 3 mmol HCl/mmol CH_5N_3 time the molar mass of CH_5N_3 divided by 4 tablets.					
14		**Wt. patient, lbs**	**Number of tablets**	**Closest whole number < 10mg/kg**			**Actual Dose, mg/kg**	
15	(a)	100.0	5.6680	5			8.821489	
16	(b)	150.0	8.5020	8			9.409588	
17	(c)	200.0	11.3360	11			9.703638	
18								
19	**Spreadsheet Documentation**		In Cells C15:C17, the number of tablets is calculated from the dosage (10 mg/kg)					
20	Cell B11=$B$5*$B$6-$B$7*$B$8		times the patient weight (lbs × kg/lb) divided by the tablet wt. This number					
21	Cell B12=$B$11*$B$2/(3*4)		is rounded off to the next lowest integer in Cells D15:D17. This is the largest					
22	Cell C15=10*B15*$B$9/$B$12		number of tablets that does not exceed the recommended dosage.					
23	Cell G15=D15*$B$12/(B15*$B$9))		The actual dosage with the integer number of tablets is calculated in Cells					
24			G15:G17 as no. tablets times wt./tablet divided by the patient weight in kg.					

14-35.

$$\%N = \frac{(24.61 \times 0.1180)\ \text{mmol HCl} \times \frac{1\ \text{mmol N}}{\text{mmol HCl}} \times \frac{0.014007\ \text{g N}}{\text{mmol N}}}{1.047\ \text{g sample}} \times 100\% = 3.885$$

14-37.

	A	B	C	D	E	F
1	**14-37 Nitrogen in a plant food preparation**					
2	Weight sample, g	0.5843				
3	Volume of HCl, mL	50.00				
4	Conc. HCl, M	0.1062				
5	Vol. NaOH, mL	11.89				
6	Conc. NaoH, M	0.0925	In Cell B8, the mmol of HCl/g sample is calculated by subtracting			
7			the mmol of NaOH used from the total mmol of HCl added and			
8	mmol HCl/g sample	7.20550	dividing by the sample weight.			
9		**Molar masses**	**Percentages**			
10	(a) N	14.007	10.09	%N		
11	(b) urea	60.06	21.64	% urea		
12	(c) $(NH_4)_2SO_4$	132.141	47.61	% $(NH_4)_2SO_4$		
13	(d) $(NH_4)_3PO_4$	149.09	35.81	% $(NH_4)_3PO_4$		
14	**Spreadsheet Documentation**					
15	Cell B8=($B$3*$B$4-$B$5*$B$6)/$B$2					
16	Cell C10=$B$8*1*B10/1000*100		The percentages are calculated in Cells C10:C13 from the no. of			
17	Cell C11=$B$8*1/2*B11/1000*100		mmol of HCl/g sample times the no. of mmol compound/mmol HC			
18	Cell C12=$B$8*1/2*B12/1000*100		times the molar mass of the compound divided by 1000 (mmolar			
19	Cell C13=$B$8*1/3*B13/1000*100		mass).			

14-39. In the first titration the analyte consumed

$$(30.00 \times 0.08421 - 10.17 \times 0.08802) \quad = \quad 1.63114 \text{ mmol HCl}$$

and

$$1.63114 \quad = \quad \text{no. mmol } NH_4NO_3 + 2 \times \text{no. mmol } (NH_4)_2SO_4$$

The amounts of the two species in the entire sample are

$$\text{no. mmol } NH_4NO_3 + 2 \times \text{no. mmol } (NH_4)_2SO_4 \quad = \quad 1.63114 \times \frac{250 \text{ mL}}{50 \text{ mL}} \qquad (1)$$

$$= \quad 8.1557$$

In the second titration the analyte consumed

$$(30.00 \times 0.08421 - 14.16 \times 0.08802) \quad = \quad 1.27994 \text{ mmol HCl}$$

and

$$1.27994 \quad = \quad 2 \times \text{no. mmol } NH_4NO_3 + 2 \times \text{no. mmol } (NH_4)_2SO_4$$

The amounts of the two species in the entire sample are

$$2 \times \text{no. mmol } NH_4NO_3 + 2 \times \text{no. mmol } (NH_4)_2SO_4 \quad = \quad 1.27994 \times \frac{250 \text{ mL}}{25 \text{ mL}} \qquad (2)$$

$$= \quad 12.7994$$

Subtracting equation (1) from (2) gives

$$\text{no. mmol } NH_4NO_3 \quad = \quad 12.7994 - 8.1557 \quad = \quad 4.6437$$

$$\text{no. mmol } (NH_4)_2SO_4 \quad = \quad \frac{12.7994 - 2 \times 4.6437}{2} \quad = \quad 1.7560$$

$$\text{percent } NH_4NO_3 \quad = \quad \frac{4.6437 \text{ mmol } NH_4NO_3 \times 0.08004 \text{ g } NH_4NO_3/\text{mmol}}{1.219 \text{ g sample}} \times 100\,\%$$

$$= \quad \underline{\underline{30.49\%}}$$

$$\text{percent } (NH_4)_2SO_4 = \frac{1.7758 \text{ mmol } (NH_4)_2SO_4 \times 0.13214 \text{ g } (NH_4)_2SO_4/\text{mmol}}{1.219 \text{ g sample}} \times 100\%$$

$$= \underline{\underline{19.04\%}}$$

14-40. For first aliquot,

$$\text{no. mmol HCl} = \text{no. mmol NaOH} + \text{no. mmol KOH} + 2 \times \text{no. mmol } K_2CO_3$$

$$\text{no. mmol KOH} + 2 \times \text{no. mmol } K_2CO_3 = 40.00 \times 0.05304 - 4.74 \times 0.04983$$
$$= 1.8854$$

For second aliquot,

$$\text{no. mmol HCl} = \text{no. mmol KOH} = 28.56 \times 0.05304 = 1.5148$$

$$\text{no. mmol } K_2CO_3 = (1.8854 - 1.5148)/2 = 0.18530$$

$$\frac{1.5148 \text{ mmol KOH} \times 0.05611 \frac{\text{g KOH}}{\text{mmol}}}{1.217 \text{ g sample} \times 50.00 \text{ mL}/500.0 \text{ mL}} \times 100\% = \underline{\underline{69.84\% \text{ KOH}}}$$

$$\frac{0.18530 \text{ mmol } K_2CO_3 \times 0.13821 \frac{\text{g } K_2CO_3}{\text{mmol}}}{1.217 \text{ g sample} \times 50.00 \text{ mL}/500.0 \text{ mL}} \times 100\% = \underline{\underline{21.04\% \; K_2CO_3}}$$

$$100\% - 69.84\% - 21.04\% = \underline{\underline{9.12\% \; H_2O}}$$

14-42.

	A	B	C	D	E
1	**14-42 Titrations with 0.06122 M HCl**				
2	M HCl	0.06122			
3	M Na_3PO_4	0.05555			
4	(a)		**mL Na_3PO_4**	**mmol base**	**mL HCl**
5	Add one proton		10.00	0.55550	9.07
6	to thymolphthalein		15.00	0.83325	13.61
7	endpoint		25.00	1.38875	22.68
8			40.00	2.22200	36.30
9			**mL Na_3PO_4**	**mmol base**	**mL HCl**
10	(b)		10.00	1.11100	18.15
11	Add two protons		15.00	1.66650	27.22
12	to bromocresol		20.00	2.22200	36.30
13	green endpoint		25.00	2.77750	45.37
14	(c)				
15	M Na_3PO_4	0.02102	**mL solution**	**mmol base**	**mL HCl**
16	M Na_2HPO_4	0.01655	20.00	1.17180	19.14
17	Add two protons		25.00	1.46475	23.93
18	to phosphate and		30.00	1.75770	28.71
19	one to mono-		40.00	2.34360	38.28
20	hydrogen phosphate				
21	(d)				
22	M NaOH	0.01655	**mL Na_3PO_4**	**mmol base**	**mL HCl**
23	Add one proton		15.00	0.56355	9.21
24			20.00	0.75140	12.27
25			35.00	1.31495	21.48
26			40.00	1.50280	24.55
27	**Spreadsheet documentation**				
28	Cell D5=C5*$B$3				
29	Cell D10=C10*$B$3*2				
30	Cell D16=$B$15*2*C16+$B$16*C16				
31	Cell D23=C23*$B$15+C23*$B$22				
32	Cell E5=D5/$B$2				
33	Cell E10=D10/$B$2				
34	Cell E16=D16/$B$2				
35	Cell E23=D23/$B$2				

	A	B	C	D	E	F
1	**14-44 Titrations of carbonate mixtures**			**Vol. to phenol., mL**	**Vol. to BCG, mL**	
2	M HCl	0.1202	(a)	22.42	22.44	
3	Volume, mL	25.00	(b)	15.67	42.13	
4	M NaOH	40.00	(c)	29.64	36.42	
5	M Na_2CO_3	105.99	(d)	16.12	32.23	
6	M $NaHCO_3$	84.01	(e)	0.00	33.33	
7	Table 14-2 gives the volume relationships in titrations of these mixtures.					
8						
9	(a) Since essentially the same volume is used for each endpoint, there is only NaOH					
10	present. We use the average volume to calculate the no. of mg NaOH/mL					
11		**mmol NaOH**	**mg NaOH/mL**			
12		2.6961	4.314			
13	(b) Since $V_{phth} < ½V_{bcg}$, only carbonate and bicarbonate are present.					
14		**mmol carbonate**	**mmol total**	**mmol bicarbonate**	**mg Na_2CO_3/ml**	**mg $NaHCO_3$/ml**
15		1.8835	5.0640	1.2970	7.985	4.358
16	(c) Now $V_{phth} > ½V_{bcg}$, so we have a mixture of NaOH and Na_2CO_3					
17		**mmol carbonate plus NaOH**	**mmol carbonate**	**mmol NaOH**	**mg Na_2CO_3/ml**	**mg NaOH/ml**
18		3.5627	0.8150	2.7478	3.455	4.396
19	(d) Since $V_{phth} = ½V_{bcg}$, we have only Na_2CO_3 present					
20		**mmol carbonate**	**mg Na_2CO_3/ml**			
21		1.9376	8.215			
22	(e) Since $V_{phth} = 0$, we have only NaHCO3 present which gains one proton.					
23		**mmol NaHCO3**	**mg $NaHCO_3$/ml**			
24		4.0063	13.46			
25	**Documentation**					
26	Cell B12=((D2+E2)/2)*$B$2		Cell D18=B18-C18			
27	Cell C12=B12*1*$B$4/$B$3		Cell E18=C18*$B$5/$B$3			
28	Cell B15=D3*$B$2		Cell F18=D18*$B$4/$B$3			
29	Cell C15=E3*$B$2		Cell B21=$D$5*$B$2			
30	Cell D15=C15-2*B15		Cell C21=B21*1*$B$5/$B$3			
31	Cell E15=B15*$B$5/$B$3		Cell B24=E6*$B$2			
32	Cell F15=D15*$B$6/$B$3		Cell C24=B24*1*$B$6/$B$3			
33	Cell B18=D4*$B$2					
34	Cell C18=($E$4-$D$4)*$B$2					

Chapter 15

15-1. **(a)** A *chelate* is a cyclic complex consisting of metal ion and a reagent that contains two or more electron donor groups located in such a position that they can bond with the metal ion to form a heterocyclic ring structure.

(c) A *ligand* is a species that contains one or more electron pair donor groups that tend to form bonds with metal ions.

(e) A *conditional formation constant* is an equilibrium constant for the reaction between a metal ion and a complexing agent that applies only when the pH and/or the concentration of other complexing ions are carefully specified.

(g) *Water hardness* is the concentration of calcium carbonate that is equivalent to the total molar concentration of all of the multivalent metal carbonates in the water.

15-2. Three general methods for performing EDTA titrations include: (1) direct titration, (2) back titration, and (3) displacement titration. Method (1) is simple, rapid, and requires but one standard reagent. Method (2) is advantageous for those metals that react so slowly with EDTA as to make direct titration inconvenient. In addition, this procedure is useful for cations for which satisfactory indicators are not available. Finally, it is useful for analyzing samples that contain anions that form sparingly soluble precipitates with the analyte under the analytical conditions. Method (3) is particularly useful in situations where no satisfactory indicators are available for direct titration.

15-4. **(a)**

$$Ag^+ + S_2O_3^{2-} \rightleftharpoons AgS_2O_3^- \qquad K_1 = \frac{[AgS_2O^-]}{[Ag^+][S_2O_3^{2-}]}$$

$$AgS_2O_3^- + S_2O_3^{2-} \rightleftharpoons Ag(S_2O_3)_2^{3-} \qquad K_2 = \frac{[Ag(S_2O_3)_2^{3-}]}{[AgS_2O_3^-][S_2O_3^{2-}]}$$

15-5. The overall formation constant β_n is equal to the product of the individual stepwise constants. Thus the overall constant for formation of $Ni(SCN)_3^-$ is

$$\beta_3 = K_1K_2K_3 = \frac{[Ni(SCN)_3^-]}{[Ni^{2+}][SCN^-]^3}$$

which is the equilibrium constant for the reaction

$$Ni^{2+} + 3SCN^- \rightleftarrows Ni(SCN)_3^-$$

and

$$\beta_2 = K_1K_2 = \frac{Ni(SCN)_2}{[Ni^{2+}][SCN^-]^2}$$

where the overall constant β_2 is for the equilibrium

$$Ni^{2+} + 2SCN^- \rightleftarrows Ni(SCN)_2$$

15-6. The Fajans determination of chloride involves a direct titration, while a Volhard approach requires two standard solutions and a filtration step to remove AgCl before back titration of the excess SCN^-.

15-9. Potassium is determined by precipitation with an excess of a standard solution of sodium tetraphenylboron. An excess of standard $AgNO_3$ is then added, which precipitates the excess tetraphenylboron ion. The excess $AgNO_3$ is then titrated with a standard solution of SCN^-. The reactions are

$$K^+ + B(C_6H_5)_4^- \rightleftarrows KB(C_6H_5)_4(s) \qquad \text{[measured excess } B(C_6H_5)_4^-\text{]}$$

$$Ag^+ + B(C_6H_5)_4^- \rightleftarrows AgB(C_6H_5)_4(s) \qquad \text{[measured excess } AgNO_3\text{]}$$

The excess $AgNO_3$ is then determined by a Volhard titration with KSCN.

15-12.

$$3.853 \text{ g reagent} \times \frac{99.7 \text{ g } Na_2H_2Y \cdot 2H_2O}{100 \text{ g reagent}} \times \frac{1 \text{ mol EDTA}}{372.24 \text{ g } Na_2H_2Y \cdot 2H_2O} \times \frac{1}{1.000 \text{ L}} =$$

$$\underline{\underline{0.01032 \text{ M EDTA}}}$$

15-14. (a)

$$26.37 \text{ mL} \times 0.0741 \frac{\text{mmol } Mg(NO_3)_2}{\text{mL}} \times \frac{1 \text{ mmol EDTA}}{\text{mmol } Mg(NO_2)_2} \times \frac{1 \text{ mL EDTA}}{0.0500 \text{ mmol EDTA}} =$$

$$\underline{\underline{39.1 \text{ mL EDTA}}}$$

(c) Letting A symbolize $CaHPO_4 \cdot 2H_2O$

$$(0.4397 \times 0.814)\text{ g A} \times \frac{1\text{ mmol A}}{0.17209\text{ g A}} \times \frac{1\text{ mmol EDTA}}{\text{mmol A}} \times \frac{1\text{ mL EDTA}}{0.0500\text{ mmol EDTA}} =$$

$$\underline{\underline{41.6\text{ mL EDTA}}}$$

(e) Proceeding as in part (c), we write

$$(0.1557 \times 0.925)\text{ g A} \times \frac{\text{mmol A}}{0.1844\text{ g}} \times \frac{2\text{ mmol EDTA}}{1\text{ mmol A}} \times \frac{1\text{ mL EDTA}}{0.0500\text{ mmol EDTA}} =$$

$$\underline{\underline{31.2\text{ mL EDTA}}}$$

15-16.

$$\frac{(21.27 \times 0.01645)\text{ mmol EDTA} \times \frac{1\text{ mmol Zn}}{\text{mmol EDTA}} \times \frac{0.06539\text{ g Zn}}{\text{mmol Zn}}}{0.7556\text{ g sample}} \times 100\% = \underline{\underline{3.028\%\text{ Zn}}}$$

15-17.

$$(15.00 \times 0.01768 - 4.30 \times 0.008120)\frac{\text{mmol Cr}}{3.00 \times 4.00\text{ cm}^2} \times \frac{51.996\text{ mg Cr}}{\text{mmol Cr}} = \underline{\underline{0.998\frac{\text{mg Cr}}{\text{cm}^2}}}$$

15-18.

$$\frac{14.77\text{ g}}{\text{L}} \times \frac{1\text{ mol AgNO}_3}{169.873\text{ g}} = 0.08695\text{ M AgNO}_3$$

(a)

$$0.2631\text{ g} \times \frac{\text{mmol NaCl}}{0.05844\text{ g}} \times \frac{1\text{ mmol AgNO}_3}{\text{mmol NaCl}} \times \frac{1\text{ mL AgNO}_3}{0.08695\text{ mmol AgNO}_3} = \underline{\underline{51.78\text{ mL AgNO}_3}}$$

(c)

$$64.13\text{ mg} \times \frac{\text{mmol Na}_3\text{AsO}_4}{207.88\text{ mg}} \times \frac{3\text{ mmol AgNO}_3}{\text{mmol Na}_3\text{AsO}_4} \times \frac{1\text{ mL}}{0.08695\text{ mmol}} = \underline{\underline{10.64\text{ mL AgNO}_3}}$$

(e)

$$25.00\text{ mL} \times \frac{0.05361\text{ mmol Na}_3\text{PO}_4}{\text{mL}} \times \frac{3\text{ mmol AgNO}_3}{\text{mmol Na}_3\text{PO}_4} \times \frac{1\text{ mL}}{0.08695\text{ mmol}} = \underline{\underline{46.24\text{ mL AgNO}_3}}$$

15-20. (a) An excess is assured if the calculation is based on a pure sample.

$$0.2513\text{ g NaCl} \times \frac{1\text{ mmol NaCl}}{0.05844\text{ g NaCl}} \times \frac{\text{mmol AgNO}_3}{\text{mmol NaCl}} \times \frac{1\text{ mL AgNO}_3}{0.09621\text{ mmol AgNO}_3} =$$

$$\underline{\underline{44.70\text{ mL AgNO}_3}}$$

(c) $$25.00\text{ mL AlCl}_3 \times \frac{0.01907\text{ mmol AlCl}_3}{\text{mL AlCl}_3} \times \frac{3\text{ mmol AgNO}_3}{\text{mmol AlCl}_3} \times \frac{1\text{ mL AgNO}_3}{0.09621\text{ mmol AgNO}_3} = \underline{\underline{14.87\text{ mL AgNO}_3}}$$

15-22. $$\frac{(13.34 \times 0.03560)\text{ mmol EDTA}}{9.76\text{ g sample}} \times \frac{1\text{ mmol Tl}_2\text{SO}_4}{2\text{ mmol EDTA}} \times \frac{0.5048\text{ g Tl}_2\text{SO}_4}{\text{mmol Tl}_2\text{SO}_4} \times 100\% = \underline{\underline{1.228\%\text{ Tl}_2\text{SO}_4}}$$

15-24.

$$\text{no. mmol Fe}^{3+} = (13.73 \times 0.01200)\text{ mmol EDTA} \times \frac{1\text{ mmol Fe}^{2+}}{\text{mmol EDTA}} = 0.16476$$

$$\text{no. mmol Fe}^{2+} = (29.62 - 13.73)\text{ mL EDTA} \times 0.01200\frac{\text{mmol EDTA}}{\text{mL EDTA}} \times \frac{1\text{ mmol Fe}^{2+}}{\text{mmol EDTA}} = 0.19068$$

$$\frac{0.16476\text{ mmol Fe}^{3+} \times 55.847\text{ mg Fe}^{3+}/\text{mmol}}{50.00\text{ mL} \times 10^{-3}\text{ L/mL}} = 184.0\frac{\text{mg Fe}^{3+}}{\text{L}} = \underline{\underline{184.0\text{ ppm Fe}^{3+}}}$$

Similarly,

$$0.19068 \times 55.847/0.05000 = \underline{\underline{213.0\text{ ppm Fe}^{2+}}}$$

15-26. no. mmol (Cd^{2+} + Pb^{2+}) = 28.89 × 0.06950 = 2.0079

no. mmol Pb^{2+} = 11.56 × 0.06950 = $\underline{0.8034}$

no. mmol Cd^{2+} = 1.2045

$$\frac{0.8034\text{ mmol Pb}^{2+} \times 0.2072\text{ g Pb/mmol Pb}}{1.509\text{ g sample} \times 50.00\text{ mL}/250.0\text{ mL}} \times 100\% = \underline{\underline{55.16\%\text{ Pb}}}$$

Similarly,

$$\frac{1.2045 \times 0.11241}{1.509 \times 50.00/250.0} \times 100 = \underline{\underline{44.86\%\text{ Cd}}}$$

15-28. $$\frac{(38.71\times 0.01294)\text{ mmol EDTA}\times\frac{1\text{ mmol ZnO}}{\text{mmol EDTA}}\times\frac{0.08139\text{ g ZnO}}{\text{mmol ZnO}}}{1.022\text{ g sample}\times 10.00\text{ mL}/250.0\text{ mL}}\times 100\,\% \;=\; \underline{\underline{99.7\%\text{ ZnO}}}$$

$$\frac{2.40\text{ mL}\times 0.002727\frac{\text{mmol ZnY}^{2-}}{\text{mL}}\times\frac{1\text{ mmol Fe}_2\text{O}_3}{2\text{ mmol ZnY}^{2-}}\times\frac{0.15969\text{ g Fe}_2\text{O}_3}{\text{mmol Fe}_2\text{O}_3}}{1.022\text{ g sample}\times 50.00\text{ mL}/250.0\text{ mL}}\times 100\,\% \;=\; \underline{\underline{0.256\%\text{ Fe}_2\text{O}_3}}$$

15-30. no. mmol (Ni + Fe + Cr) in sample =

$(50.00 \times 0.05182 - 5.11 \times 0.06241) \times 250.0/50.00 = 11.360$

no. mmol (Ni + Fe) in sample =

$(36.28 \times 0.05182) \times 250.0/50.00 = 9.4001$

no. mmol Cr in sample = $11.360 - 9.4001 = 1.9603$

no. mmol Ni in sample = $25.91 \times 0.05182 \times 250/50.00 = 6.7133$

no. mmol Fe in sample = $9.4001 - 6.7133 = 2.6868$

% Cr = $(1.9603 \times 0.051996/0.6472) \times 100\,\%$ = <u>15.75%</u>

% Ni = $(6.7133 \times 0.05869/0.6472) \times 100\,\%$ = <u>60.88%</u>

% Fe = $(2.6868 \times 0.055847/0.6472) \times 100\,\%$ = <u>23.18%</u>

15-32.

	A	B	C	D	E	F	G	H	I
1	**15-32 Conditional constants for Fe^{2+}- EDTA complex**								
2	Note: The conditional constant K'_{MY} is the product of α_4 and K_{MY} (Equation 15-23).								
3	The value of K_{MY} is found in Table 15-5.								
4	K_{MY}	2.10E+14	**pH**	**D**	α_4	K'_{MY}			
5	K_1	1.02E-02	6.0	3.69E-17	2.25E-05	4.7E+09	Note that Excel does not		
6	K_2	2.14E-03	8.0	1.54E-19	5.39E-03	1.1E+12	follow the rounding rules		
7	K_3	6.92E-07	10.0	2.34E-21	3.55E-01	7.5E+13	developed in Section 6D-3.		
8	K_4	5.50E-11							
9	**Documentation**								
10	Cell D5=(10^-C5)^4+$B$5*(10^-C5)^3+$B$5*$B$6*(10^-C5)^2+$B$5*$B$6*$B$7*(10^-C5)+								
11							$B$5*$B$6*$B$7*$B$8		
12	Cell E5=$B$5*$B$6*$B$7*$B$8/D5								
13	Cell F5=E5*$B$4								

15-34. First we calculate K'_{MY} (cell F5) by Equation 15-25 and note that

$$K'_{MY} = \frac{[SrY^{2-}]}{[Sr^{2+}]\,c_T}$$

0.00 mL

The initial $[Sr^{2+}]$ is equal to the analytical concentration $c_{Sr^{2+}}$ of Sr^{2+}.

Pre-equivalence point region

We calculate $c_{Sr^{2+}}$ and $c_{SrY^{2-}}$ from

$$c_{Sr^{2+}} = \frac{\text{initial mmol } Sr^{2+} - \text{mmol EDTA added}}{\text{total solution volume, mL}}$$

$$c_{SrY^{2-}} = \frac{\text{mmol EDTA added}}{\text{total solution volume, mL}}$$

Here the equilibrium concentration of Sr^{2+} approximately equals the analytical concentration, $c_{Sr^{2+}}$.

Equivalence Point

At **25.00 mL,** $[SrY^{2-}] \approx c_{SrY^{2-}}$ and $[Sr^{2+}] \approx c_T$

From above,

$$K'_{MY} = \frac{[SrY^{2-}]}{[Sr^{2+}]\,c_T} \approx \frac{[SrY^{2-}]}{[Sr^{2+}]^2}$$

$$[Sr^{2+}] = \sqrt{\frac{[SrY^{2-}]}{K'_{MY}}}$$

Beyond the equivalence point

Here,

$$c_T = \frac{\text{mmol EDTA added} - \text{mmol } Sr^{2+} \text{ initially present}}{\text{total solution volume, mL}}$$

$$[SrY^{2-}] = \frac{\text{mmol } Sr^{2+} \text{ initially present}}{\text{total solution volume, mL}}$$

$$[Sr^{2+}] = \frac{[SrY^{2-}]}{c_T - K'_{MY}}$$

The plot of pSr versus the volume of EDTA added is shown in the spreadsheet.

	A	B	C	D	E	F	G	H
1	**15-34 Titration of 50.00 mL of 0.01000 M Sr^{2+} with 0.02000 M EDTA**							
2	Note: The conditional constant K'_{MY} is the product of α_4 and K_{MY} (Equation 15-23).							
3	The value of K_{MY} is found in Table 15-5. See Problem 15-32 for calculation of K'.							
4	K_{MY}	4.30E+08	**pH**	**D**	α_4	K'_{MY}		
5	EDTA K_1	1.02E-02	11.0	9.82E-22	8.46E-01	3.64E+08		
6	K_2	2.14E-03						
7	K_3	6.92E-07						
8	K_4	5.50E-11						
9	Initial conc. Sr^{2+}	0.01000	Initial Vol.	50.00				
10	Initial conc. EDTA	0.02000						
11	**Vol. EDTA, mL**	c_{Sr2+}	c_{SrY2-}	c_T	**[Sr^{2+}]**	**[SrY^{2-}]**		**pSr**
12	0.00	0.01000	0		0.01000			2.00
13	10.00	0.00500	0.00333		0.00500			2.30
14	24.00	0.00027	0.00649		0.00027			3.57
15	24.90	0.00003	0.00665		0.00003			4.57
16	25.00	0.00000	0.00667	0.00667	4.28E-06	0.00667		5.37
17	25.10		0.00666	2.66E-05	6.87E-07	0.00666		6.16
18	26.00		0.00658	2.63E-04	6.87E-08	0.00658		7.16
19	30.00		0.00625	1.25E-03	1.37E-08	0.00625		7.86
20	**Documentation**							
21	Cell B12=($B$9*$D$9-$B$10*A12)/($D$9+A12)							
22	Cell C12=($B$10*A12)/($D$9+A12)							
23	Cell C16=($B$10*$A$16)/($D$9+A16)							
24	Cell D17=($B$10*A17-$D$9*$B$9)/($D$9+A17)							
25	Cell D16=($B$10*$A$16)/($D$9+A16)							
26	Cell E12=B12							
27	Cell E16=SQRT(C16/$F$5)							
28	Cell E17=C17/(D17*$F$5)							
29	Cell F16=C16							
30	Cell H12=-LOG10(E12)							
31								
32								
33								
34								
35								
36								
37								
38								
39								
40								
41								
42								
43								
44								

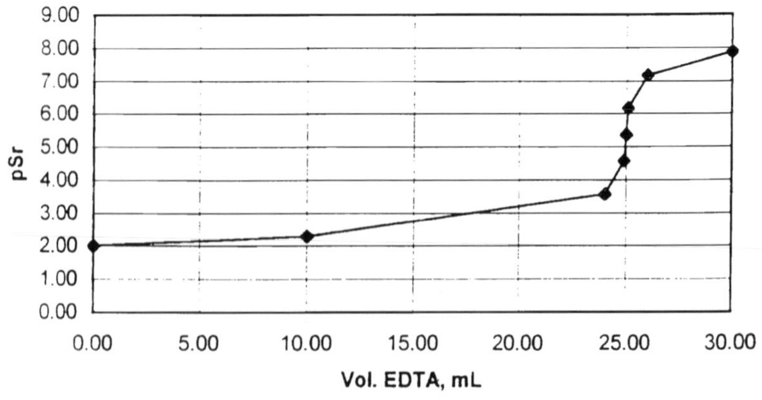

15-36. (a) There are $25.00 \times 0.0500 = 1.25$ mmol Ag^+ initially present. At the equivalence point,

$$V_{SCN^-} = \frac{1.25 \text{ mmol } SCN^-}{0.0250 \text{ mmol/mL}} = 50.00 \text{ mL}$$

Pre-equivalence point region

$$c_{AgNO_3} = [Ag^+] = \frac{\text{initial no. of mmol } AgNO_3 - \text{ mmol } NH_4SCN \text{ added}}{\text{total volume of solution, mL}}$$

$pAg = -\log[Ag^+]$ The results are shown in the spredsheet for part (a).

$[SCN^-] = K_{sp}/[Ag^+]$

Equivalence point

Here $[Ag^+] = [SCN^-] = \sqrt{K_{sp}}$. For the result see the spreadsheet for this part.

Post-equivalence point

Here we calculate the SCN^- concentration and obtain the Ag^+ concentration from K_{sp}.

$$c_{NH_4SCN} \approx [SCN^-] = \frac{\text{mmol } NH_4SCN \text{ added} - \text{mmol } AgNO_3 \text{ initially present}}{\text{total solution volume in mL}}$$

$$[Ag^+] = K_{sp}/[SCN^-]$$

(c) The method is identical to that in (a). The results are shown in the spreadsheet for these parts.

(e) An identical method to that in part (a) is used. The results are shown in the spreadsheets.

15-36. (a)

	A	B	C	D	E	F	G	H
1	**15-36(a) Titration of 25.00 mL of 0.05000 M $AgNO_3$ with 0.02500 M NH_4SCN**							
2	K_{sp}	1.10E-12						
3	Initial conc. Ag^+	0.05000						
4	Conc. SCN^-	0.02500						
5	Initial volume	25.00						
6								
7								
8	**Vol. SCN^-, mL**	**c_{AgNO3}**	**c_{NH4SCN}**	**$[Ag^+]$**	**$[SCN^-]$**	**pAg**		
9	30.00	0.00909		0.00909	1.21E-10	2.04		
10	40.00	0.00385		0.00385	2.86E-10	2.41		
11	49.00	0.00034		0.00034	3.26E-09	3.47		
12	50.00	0.00000		1.05E-06	1.05E-06	5.98		
13	51.00		0.00033	3.34E-09	0.00033	8.48		
14	60.00		0.00294	3.74E-10	0.00294	9.43		
15	70.00		0.00526	2.09E-10	0.00526	9.68		
16								
17	**Documentation**							
18	Cell B9=($B$3*$B$5-$B$4*A9)/($B$5+A9)				**Note in cell F10 that Excel does not**			
19	Cell D9=B9				**round this value up as our rules dictate.**			
20	Cell E9=$B$2/D9				**We would say pAg = 2.42.**			
21	Cell F9=-LOG10(D9)							
22	Cell D12=SQRT($B$2)							
23	Cell E12=SQRT($B$2)							
24	Cell C13=($B$4*A13-$B$3*$B$5)/($B$5+A13)							
25	Cell E13=C13							
26	Cell D13=$B$2/E13							
27								
28								
29								
30								
31								
32								
33								
34								
35								
36								
37								
38								
39								
40								
41								
42								
43								

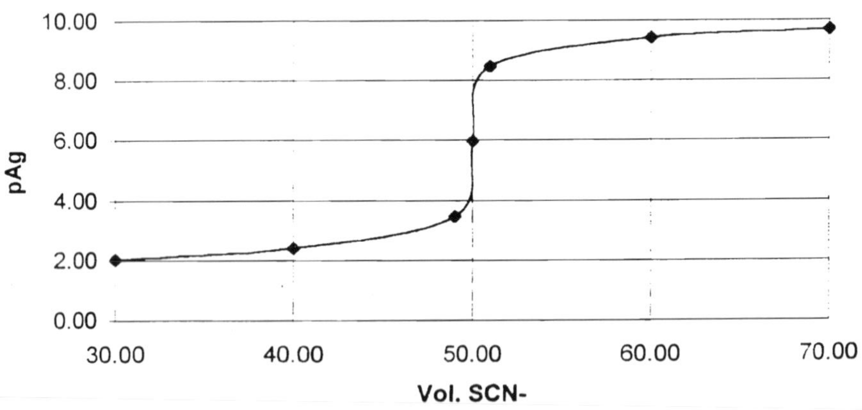
10.00
8.00
6.00
4.00
2.00
0.00
pAg
30.00
40.00
50.00
60.00
70.00
Vol. SCN-

	A	B	C	D	E	F	G	H	I
1	**15-36(b) and (c) Titrations of $AgNO_3$ with KI and NaCl**								
2	For AgI K_{sp}	8.30E-17		In (b) the equivalence point is at 40.00 mL, while in (c)					
3	For AgCl K_{sp}	1.82E-10		it occurs at 30.00 mL					
4	(b) Initial conc. Ag^+	0.06000							
5	(c) Initial conc. Ag^+	0.07500							
6	(b) Conc. I^-	0.03000							
7	(c) Conc. Cl^-	0.07500							
8	(b) Initial volume	20.00							
9	(c) Initial volume	30.00							
10	(b)								
11	**Vol. I^-, mL**	c_{AgNO3}	c_{KI}	**$[Ag^+]$**	**$[I^-]$**	**pAg**			
12	20.00	0.01500		0.01500	5.53E-15	1.82			
13	30.00	0.00600		0.00600	1.38E-14	2.22			
14	39.00	0.00051		0.00051	1.63E-13	3.29			
15	40.00	0.00000		9.11E-09	9.11E-09	8.04			
16	41.00		0.00049	1.69E-13	0.00049	12.77			
17	50.00		0.00429	1.94E-14	0.00429	13.71			
18	60.00		0.00750	1.11E-14	0.00750	13.96			
19	(c)								
20	**Vol. Cl^-, mL**	c_{AgNO3}	c_{NaCl}	**$[Ag^+]$**	**$[Cl^-]$**	**pAg**			
21	10.00	0.03750		0.03750	4.85E-09	1.43			
22	20.00	0.01500		0.01500	1.21E-08	1.82			
23	29.00	0.00127		0.00127	1.43E-07	2.90			
24	30.00	0.00000		1.35E-05	1.35E-05	4.87			
25	31.00		0.00123	1.48E-07	0.00123	6.83			
26	40.00		0.01071	1.70E-08	0.01071	7.77			
27	50.00		0.01875	9.71E-09	0.01875	8.01			
28	**Documentation**								
29	Cell B12=($B$4*$B$8-$B$6*A12)/($B$8+A12)				Cell B21=($B$5*$B$9-$B$7*A21)/($B$9+A21)				
30	Cell D12=B12				Cell D21=B21				
31	Cell E12=$B$2/D12				Cell E21=$B$3/D21				
32	Cell F12=-LOG10(D12)				Cell F21=-LOG10(D21)				
33	Cell D15=E15=SQRT($B$2)				Cell D24=E24=SQRT($B$3)				
34	Cell C16=($B$6*A16-$B$4*$B$8)/($B$8+A16)				Cell C25=($B$7*A25-$B$5*$B$9)/($B$9+A25)				
35	Cell E16=C16				Cell E25=C25				
36	Cell D16=$B$2/E16				Cell D25=$B$3/E25				

15-36. (d), (e)

	A	B	C	D	E	F	G	H	I
1	**15-36(d) and (e) Titrations forming insoluble sulfates**								
2	For $PbSO_4$ K_{sp}	1.60E-08		In (d) the equivalence point is at 70.00 mL. while in (e)					
3	For $BaSO_4$, K_{sp}	1.10E-10		it occurs at 20.00 mL					
4	(d) Initial conc.SO_4^{2-}	0.40000							
5	(e) Initial conc. Ba^{2+}	0.02500							
6	(d) Conc. Pb^{2+}	0.20000							
7	(e) Conc. SO_4^{2-}	0.05000							
8	(d) Initial volume	35.00							
9	(e) Initial volume	40.00							
10	(d)								
11	**Vol. Pb^{2+}, mL**	c_{Na2SO4}	$c_{Pb(No3)2}$	**$[SO_4^{2-}]$**	**$[Pb^{2+}]$**	**pPb**			
12	50.00	0.04706		0.04706	3.40E-07	6.47			
13	60.00	0.02105		0.02105	7.60E-07	6.12			
14	69.00	0.00192		0.00192	8.32E-06	5.08			
15	70.00	0.00000		1.26E-04	1.26E-04	3.90			
16	71.00		0.00189	8.48E-06	0.00189	2.72			
17	80.00		0.01739	9.20E-07	0.01739	1.76			
18	90.00		0.03200	5.00E-07	0.03200	1.49			
19	(e)								
20	**Vol. SO_4^{2}, mL**	c_{BaCl2}	c_{Na2SO4}	**$[SO_4^{2-}]$**	**$[Ba^{2+}]$**	**pBa**			
21	0.00	0.02500		0.02500	4.40E-09	1.60			
22	10.00	0.01000		0.01000	1.10E-08	2.00			
23	19.00	0.00085		0.00085	1.30E-07	3.07			
24	20.00	0.00000		1.05E-05	1.05E-05	4.98			
25	21.00		0.00082	1.34E-07	0.00082	6.87			
26	30.00		0.00714	1.54E-08	0.00714	7.81			
27	40.00		0.01250	8.80E-09	0.01250	8.06			
28	**Documentation**								
29	Cell B12=($B$4*$B$8-$B$6*A12)/($B$8+A12)				Cell B21=($B$5*$B$9-$B$7*A21)/($B$9+A21)				
30	Cell D12=B12				Cell D21=B21				
31	Cell E12=$B$2/D12				Cell E21=$B$3/D21				
32	Cell F12=-LOG10(D12)				Cell F21=-LOG10(D21)				
33	Cell D15=E15=SQRT($B$2)				Cell D24=E24=SQRT($B$3)				
34	Cell C16=($B$6*A16-$B$4*$B$8)/($B$8+A16)				Cell C25=($B$7*A25-$B$5*$B$9)/($B$9+A25)				
35	Cell E16=C16				Cell E25=C25				
36	Cell D16=$B$2/E16				Cell D25=$B$3/E25				

Chapter 16

16-1. **(a)** *Oxidation* is a process in which a species loses one or more electrons.

(c) A *salt bridge* is a device that provides electrical contact but prevents mixing of dissimilar solutions in an electrochemical cell.

(e) The *Nernst equation* relates the potential to the concentrations (strictly, activities) of the participants in an electrochemical half-cell.

16-2. **(a)** The *electrode potential* is the potential of an electrochemical cell in which a standard hydrogen electrode acts as the reference electrode on the left and the half-cell of interest is the electrode on the right.

(c) The *standard electrode potential* for a half-reaction is the potential of a *cell* consisting of the half-reaction of interest on the right and a standard hydrogen electrode behaving as the reference on the left. The activities of all of the participants in the half-reaction are specified as having a value of unity. The additional specification that the standard hydrogen electrode is the left-hand electrode implies that the standard potential for a half-reaction is always a *reduction potential.*

(e) An *oxidation potential* is the potential of an electrochemical cell in which the right-hand electrode is a standard hydrogen electrode and the half-cell of interest is on the left.

16-3. **(a)** *Reduction* is the process whereby a substance acquires electrons; a *reducing agent* is a supplier of electrons.

(c) The *anode* of an electrochemical cell is the electrode at which oxidation occurs. The *cathode* is the electrode at which reduction occurs.

(e) The *standard electrode potential* is the potential of an electrochemical cell in which the standard hydrogen electrode acts as the reference electrode on the left and all participants in the right-hand electrode process have unit activity. The formal potential differs in that the molar *concentrations* of the reactants and products are unity and the concentration of other species in the solution are carefully specified.

16-4. The first standard potential is for a solution that is saturated with I_2, which has an $I_2(aq)$ activity significantly less than one. The second potential is for a *hypothetical* half-cell in which the $I_2(aq)$ activity is unity. Such a half-cell, if it existed, would have a greater potential since the driving force for the reduction would be greater at the higher I_2 concentration. The second half-cell potential, although hypothetical, is nevertheless useful for calculating electrode potentials for solutions that are undersaturated in I_2.

16-5. It is necessary to bubble hydrogen through the electrolyte in a hydrogen electrode in order to keep the solution saturated with the gas. Only under these circumstances is the hydrogen activity constant so that the electrode potential is constant and reproducible.

16-7. **(a)** $2Fe^{3+} + Sn^{2+} \rightarrow 2Fe^{2+} + Sn^{4+}$

(c) $2NO_3^- + Cu(s) + 4H^+ \rightarrow 2NO_2(g) + 2H_2O + Cu^{2+}$

(e) $Ti^{3+} + Fe(CN)_6^{3-} + H_2O \rightarrow TiO^{2+} + Fe(CN)_6^{4-} + 2H^+$

(g) $2Ag(s) + 2I^- + Sn^{4+} \rightarrow 2AgI(s) + Sn^{2+}$

(i) $5HNO_2 + 2MnO_4^- + H^+ \rightarrow 5NO_3^- + 2Mn^{2+} + 3H_2O$

16-8. **(a)** Oxidizing Agent Fe^{3+}; $Fe^{3+} + e^- \rightleftarrows Fe^{2+}$

Reducing Agent Sn^{2+}; $Sn^{2+} \rightleftarrows Sn^{4+} + 2e^-$

(b) Oxidizing Agent Ag^+; $Ag^+ + e^- \rightleftarrows Ag(s)$

Reducing Agent Cr; $Cr(s) \rightleftarrows Cr^{3+} + 3e^-$

(c) Oxidizing Agent NO_3^-; $NO_3^- + 2H^+ + e^- \rightleftarrows NO_2(g) + H_2O$

Reducing Agent Cu; $Cu(s) \rightleftarrows Cu^{2+} + 2e^-$

(d) Oxidizing Agent MnO_4^-; $MnO_4^- + 8H^+ + 5e^- \rightleftarrows Mn^{2+} + 4H_2O$

Reducing Agent H_2SO_3; $H_2SO_3 + H_2O \rightleftarrows SO_4^{2-} + 4H^+ + 2e^-$

(e) Oxidizing Agent $Fe(CN)_6^{3-}$; $Fe(CN)_6^{3-} + e^- \rightleftarrows Fe(CN)_6^{4-}$

Reducing Agent Ti^{3+}; $Ti^{3+} + H_2O \rightleftarrows TiO^{2+} + 2H^+ + e^-$

(f) Oxidizing Agent Ce^{4+}; $Cu^{4+} + e^- \rightleftarrows Ce^{3+}$

Reducing Agent H_2O_2; $H_2O_2 \rightleftarrows O_2(g) + 2H^+ + 2e^-$

(g) Oxidizing Agent Sn^{4+}; $Sn^{4+} + 2e^- \rightleftarrows Sn^{2+}$

Reducing Agent Ag; $Ag(s) + I^- \rightleftarrows AgI(s) + e^-$

(h) Oxidizing Agent UO_2^{2+}; $UO_2^{2+} + 4H^+ + 2e^- \rightleftarrows U^{4+} + 2H_2O$

Reducing Agent Zn; $Zn(s) \rightleftarrows Zn^{2+} + 2e^-$

(i) Oxidizing Agent MnO_4^-; $MnO_4^- + 8H^+ + 5e^- \rightleftarrows Mn^{2+} + 4H_2O$

Reducing Agent HNO_2; $HNO_2 + H_2O \rightleftarrows NO_3^- + 3H^+ + 2e^-$

(j) Oxidizing Agent IO_3^-; $IO_3^- + 2Cl^- + 6H^+ + 4e^- \rightleftarrows ICl_2^- + 3H_2O$

Reducing Agent H_2NNH_2; $N_2H_4 \rightleftarrows N_2(g) + 4H^+ + 4e^-$

16-9. (a) $MnO_4^- + 5VO^{2+} + 11H_2O \rightarrow Mn^{2+} + 5V(OH)_4^+ + 2H^+$

(c) $Cr_2O_7^{2-} + 3U^{4+} + 2H^+ \rightarrow 2Cr^{3+} + 3UO_2^{2+} + H_2O$

(e) $IO_3^- + 5I^- + 6H^+ \rightarrow 3I_2 + H_2O$

(g) $HPO_3^{2-} + 2MnO_4^- + 3OH^- \rightarrow PO_4^{3-} + 2MnO_4^{2-} + 2H_2O$

(i) $V^{2+} + 2V(OH)_4^+ + 2H^+ \rightarrow 3VO^{2+} + 5H_2O$

16-11. (a)

$AgBr(s) + e^- \rightleftarrows Ag(s) + Br^-$

$V^{2+} \rightleftarrows V^{3+} + e^-$

$Tl^{3+} + 2e^- \rightleftarrows Tl^+$

$Fe(CN)_6^{4-} \rightleftarrows Fe(CN)_6^{3-} + e^-$

$V^{3+} + e^- \rightleftarrows V^{2+}$

$Zn \rightleftarrows Zn^{2+} + 2e^-$

$Fe(CN)_6^{3-} + e^- \rightleftarrows Fe(CN)_6^{4-}$

$Ag(s) + Br^- \rightleftarrows AgBr(s) + e^-$

$S_2O_8^{2-} + 2e^- \rightleftarrows 2SO_4^{2-}$

$Tl^+ \rightleftarrows Tl^{3+} + 2e^-$

(b), (c)

	E^0
$S_2O_8^{2-} + 2e^- \rightleftarrows 2SO_4^{2-}$	2.01
$Tl^{3+} + 2e^- \rightleftarrows Tl^+$	1.25
$Fe(CN)_6^{3-} + e^- \rightleftarrows Fe(CN)_6^{4-}$	0.36
$AgBr(s) + e^- \rightleftarrows Ag(s) + Br^-$	0.073
$V^{3+} + e^- \rightleftarrows V^{2+}$	-0.256
$Zn^{2+} + 2e^- \rightleftarrows Zn(s)$	-0.763

16-13. (a)

$$E_{Cu} = 0.337 - \frac{0.0592}{2}\log\frac{1}{0.0440} = \underline{\underline{0.297\text{ V}}}$$

(b) $K_{sp}\text{CuCl} = 1.9 \times 10^{-7} = [Cu^+][Cl^-]$

$$E_{Cu} = 0.521 - \frac{0.0592}{1}\log\frac{1}{[Cu^+]} = 0.521 - \frac{0.0591}{1}\log\frac{[Cl^-]}{K_{sp}}$$

$$= 0.521 - \frac{0.0592}{1}\log\frac{0.0750}{1.9 \times 10^{-7}} = 0.521 - \frac{0.0592}{1}\log 3.95 \times 10^5$$

$$= 0.521 - 0.331 = \underline{\underline{0.190\text{ V}}}$$

(c)

$$K_{sp}\,Cu(OH)_2 = 4.8\times10^{-20} \qquad [Cu^{2+}] = \frac{4.8\times10^{-20}}{[OH^-]^2}$$

$$E_{Cu} = 0.337 - \frac{0.0592}{2}\log\frac{[OH^-]^2}{K_{sp}} = 0.337 - \frac{0.0592}{2}\log\frac{(0.0400)^2}{4.8\times10^{-20}}$$

$$= 0.337 - \frac{0.0592}{2}\log 3.33\times10^{16} = 0.337 - 0.489 = \underline{\underline{-0.152\text{ V}}}$$

(d)

$$\beta_4 = 5.62\times10^{11} = \frac{[Cu(NH_3)_4^{2+}]}{[Cu^{2+}][NH_3]^4}$$

$$\frac{1}{[Cu^{2+}]} = \frac{(5.62\times10^{11})(0.128)^4}{0.0250} = 6.03\times10^9$$

$$E_{Cu} = 0.337 - \frac{0.0592}{2}\log 6.03\times10^9 = 0.337 - 0.289 = \underline{\underline{0.048\text{ V}}}$$

(e) Substituting into Equation 15-23 gives

$$\frac{[CuY^{2-}]}{[Cu^{2+}]\,c_T} = \alpha_4 K_{CuY} = 3.6\times10^{-9}\times6.3\times10^{18} = 2.27\times10^{10}$$

$$[CuY^{2-}] \approx 4.00\times10^{-3}$$

$$c_T = 2.90\times10^{-2} - 4.00\times10^{-3} = 2.50\times10^{-2}$$

$$\frac{4.00\times10^{-3}}{[Cu^{2+}]\,2.50\times10^{-2}} = 2.27\times10^{10}$$

$$\frac{1}{[Cu^{2+}]} = \frac{2.27\times10^{10}\times2.50\times10^{-2}}{4.00\times10^{-3}} = 1.42\times10^{11}$$

$$E = 0.337 - \frac{0.0592}{2}\log\frac{1}{[Cu^{2+}]} = 0.337 - \frac{0.0592}{2}\log 1.42\times10^{11}$$

$$= 0.337 - 0.330 = \underline{\underline{0.007\text{ V}}}$$

16-16. $PtCl_4^{2-} + 2e^- \rightleftarrows Pt(s) + 4Cl^- \qquad E^0 = 0.73\text{ V}$

(a)
$$E_{Pt} = 0.73 - \frac{0.0592}{2}\log\frac{(0.1492)^4}{0.0263} = 0.73 - (-0.051) = \underline{\underline{0.78\text{ V}}}$$

(b)
$$E_{Pt} = 0.154 - \frac{0.0592}{2}\log\frac{2.50\times10^{-3}}{7.50\times10^{-2}} = 0.154 - (-0.044) = \underline{\underline{0.198\text{ V}}}$$

(c)
$$E_{Pt} = 0.000 - \frac{0.0592}{2}\log\frac{1.00}{(1.00\times10^{-6})^2} = \underline{\underline{-0.355\text{ V}}}$$

(d) $VO^{2+} + 2H^+ + e^- \rightleftarrows V^{3+} + H_2O \qquad E^0 = \underline{\underline{0.359\text{ V}}}$

$$E_{Pt} = 0.359 - \frac{0.0592}{1}\log\frac{0.0586\times2}{(0.0353)(0.100)^2} = 0.359 - 0.149 = \underline{\underline{0.210\text{ V}}}$$

(e) $2Fe^{3+} + Sn^{2+} \rightarrow 2Fe^{2+} + Sn^{4+}$

no. mmol Sn^{2+} taken $= 25.00\times0.0918 = 2.295$

no. mmol Fe^{3+} taken $= 25.00\times0.1568 = 3.920$

no. mmol Sn^{4+} formed $= 3.920\text{ mmol }Fe^{2+}\times\dfrac{1\text{ mmol }Sn^{4+}}{2\text{ mmol }Fe^{3+}} = 1.960$

no. mmol Sn^{2+} remaining $= 2.295 - 1.960 = 0.335$

$$E_{Pt} = 0.154 - \frac{0.0592}{2}\log\frac{0.335/50.0}{1.96/50.0} = 0.154 - (-0.023) = \underline{\underline{0.177\text{ V}}}$$

(f) $V(OH)_4^+ + V^{3+} \rightarrow 2VO^{2+} + 2H_2O$

no. mmol $V(OH)_4^+$ taken $= 25.00\times0.0832 = 2.080$

no. mmol V^{3+} taken $= 2\times50.00\times0.01087 = 1.087$

no. mmol VO^{2+} formed $= 2\times1.087 = 2.174$

$$\text{no. mmol } V(OH)_4^+ \text{ remaining} = 2.080 - 1.087 = 0.993$$

$$E_{Pt} = 1.00 - 0.0592 \log \frac{2.174/75.00}{(0.993/75.00)(0.1000)^2} = 1.00 - 0.139 = \underline{\underline{0.86 \text{ V}}}$$

16-18. (a)
$$E_{Ni} = -0.250 - \frac{0.0592}{2} \log \frac{1}{0.0943} = -0.250 - 0.030 = \underline{\underline{-0.280 \quad \text{anode}}}$$

(b)
$$E_{Ag} = -0.151 - \frac{0.0592}{1} \log 0.0922 = -0.151 + 0.061 = \underline{\underline{-0.090 \text{ V} \quad \text{anode}}}$$

(c)
$$E_{Pt} = 1.229 - \frac{0.0592}{4} \log \frac{1}{(780/760)(1.50 \times 10^{-4})^4} = 1.229 - 0.226 =$$

$$\underline{\underline{1.003 \text{ V} \quad \text{cathode}}}$$

(d)
$$E_{Pt} = 0.154 - \frac{0.0592}{2} \log \frac{0.0944}{0.350} = 0.154 + 0.017 = \underline{\underline{0.171 \text{ V} \quad \text{cathode}}}$$

(e)
$$E_{Ag} = 0.017 - \frac{0.0592}{1} \log \frac{(0.1439)^2}{0.00753} = 0.017 - 0.026 = \underline{\underline{-0.009 \text{ V} \quad \text{anode}}}$$

16-20. $2Ag^+ + 2e^- \rightleftarrows 2Ag(s) \qquad E^0 = 0.799 \text{ V}$

$$[Ag^+]^2[SO_3^{2-}] = 1.5 \times 10^{-14} = K_{sp}$$

$$E = 0.799 - \frac{0.0592}{2} \log \frac{1}{[Ag^{2+}]^2} = 0.799 - \frac{0.0592}{2} \log \frac{[SO_3^{2-}]}{K_{sp}}$$

When $[SO_3^{2-}] = 1.00$, $\quad E = E^0$ for $Ag_2SO_3(s) + 2e^- \rightleftarrows 2Ag(s) + SO_3^{2-}$

Thus,

$$E^0 = 0.799 - 0.0296 \log (1.00/K_{sp}) = 0.799 - 0.0296 \log (1/1.5 \times 10^{-14})$$

$$= 0.799 - 0.409 = \underline{\underline{0.390 \text{ V}}}$$

16-22. $2Tl^+ + 2e^- \rightleftarrows 2Tl \qquad E^0 = -0.336\ V$

$$[Tl^+]^2[S^{2-}] = 6\times10^{-22}$$

$$E = -0.336 - \frac{0.0592}{2}\log\frac{1}{[Tl^+]^2} = -0.336 - \frac{0.0592}{2}\log\frac{[S^{2-}]}{K_{sp}}$$

When $[S^{2-}] = 1.00$,

$$E = E^0_{Tl_2S} = -0.336 - \frac{0.0592}{2}\log\frac{1}{6\times10^{-22}} = -0.96\ V$$

16-24.

$$E = -0.763 - 0.0296\log\frac{1}{[Zn^{2+}]}$$

$$\frac{[ZnY^{2-}]}{[Y^{4-}][Zn^{2+}]} = 3.2\times10^{16} \quad \text{or} \quad [Zn^{2+}] = \frac{[ZnY^{2-}]}{[Y^{4-}]\,3.2\times10^{16}}$$

$$E = -0.763 - 0.0296\log\frac{[Y^{4-}]\,3.2\times10^{16}}{[ZnY^{2-}]}$$

When $[Y^{4-}] = [ZnY^{2-}] = 1.00, \qquad E = E^0_{ZnY^{2-}}$

$$E^0_{ZnY^{2-}} = -0.763 - 0.0296\log(1.00\times3.2\times10^{16}/1.00) = \underline{\underline{-1.25\ V}}$$

16-25.

$$[Fe^{3+}] = \frac{[FeY^-]}{1.3\times10^{25}[Y^{4-}]} \quad \text{and} \quad [Fe^{2+}] = \frac{[FeY^{2-}]}{2.1\times10^{14}[Y^{4-}]}$$

$$E = 0.771 - 0.0592\log\frac{[Fe^{2+}]}{[Fe^{3+}]}$$

$$= 0.771 - 0.0592\log\frac{[FeY^{2-}]\,1.3\times10^{25}[Y^{4-}]}{2.1\times10^{14}[Y^{4-}][FeY^-]}$$

$E = E^0_{FeY^-}$ when $[FeY^{2-}]$, $[FeY^-]$, and $[Y^{4-}] = 1.000$

$$E^0 = 0.771 - 0.0592 \log (1.3 \times 10^{25} / 2.1 \times 10^{14})$$

$$= 0.771 - 0.639 = \underline{\underline{0.13 \text{ V}}}$$

Chapter 17

17-1. The electrode potential of a system is the electrode potential of all half-cell processes at equilibrium in the system.

17-2. **(a)** *Equilibrium* is the state that a system assumes after each addition of reagent. *Equivalence* refers to a particular equilibrium state when a stoichiometric amount of titrant has been added.

17-4. For points before equivalence, potential data are computed from the analyte standard potential and the analytical conentrations of the analyte and its reaction product. Post-equivalence point data are based upon the standard potential for the titrant and its analytical concentrations. The equivalence point potential is computed from the two standard potentials and the stoichiometric relation between the analyte and titrant.

17-6. An asymmetric titration curve will be encountered whenever the titrant and the analyte react in a ratio that is not 1:1.

17-8. **(a)**

$$E_{right} = -0.277 - \frac{0.0592}{2}\log\frac{1}{6.78\times10^{-3}} = -0.341\text{ V}$$

$$E_{left} = -0.763 - \frac{0.0592}{2}\log\frac{1}{0.0955} = -0.793\text{ V}$$

$$E_{cell} = E_{right} - E_{left} = -0.341 - (-0.793) = \underline{\underline{0.452\text{ V}}}$$

Oxidation on the left, reduction on the right.

(b)

$$E_{right} = 0.854 - \frac{0.0592}{2}\log\frac{1}{0.0671} = 0.819\text{ V}$$

$$E_{left} = 0.771 - 0.0592\log\frac{0.0681}{0.1310} = 0.788\text{ V}$$

$$E_{cell} = E_{right} - E_{left} = 0.819 - 0.788 = \underline{\underline{0.031\text{ V}}}$$

Oxidation on the left, reduction on the right.

(c)

$$E_{\text{right}} = 1.229 - \frac{0.0592}{4}\log\frac{1}{1.12\,(0.0794)^4} = 1.165\text{ V}$$

$$E_{\text{left}} = 0.799 - 0.0592\log\frac{1}{0.1544} = 0.751\text{ V}$$

$$E_{\text{cell}} = E_{\text{right}} - E_{\text{left}} = 1.165 - 0.751 = \underline{\underline{0.414\text{ V}}}$$

Oxidation on the left, reduction on the right.

(d)

$$E_{\text{right}} = -0.151 - 0.0592\log 0.1350 = -0.100\text{ V}$$

$$E_{\text{left}} = 0.337 - \frac{0.0592}{2}\log\frac{1}{0.0601} = 0.301\text{ V}$$

$$E_{\text{cell}} = E_{\text{right}} - E_{\text{left}} = -0.100 - 0.301 = \underline{\underline{-0.401\text{ V}}}$$

Reduction on the left, oxidation on the right.

(e)

$$\frac{[H_3O^+][HCOO^-]}{[HCOOH]} = 1.80\times10^{-4} = \frac{[H_3O^+]\times0.0764}{0.1302}$$

$$[H_3O^+] = 1.80\times10^{-4}\times0.1302/0.0764 = 3.07\times10^{-4}$$

$$E_{\text{right}} = 0.000 - \frac{0.0592}{2}\log\frac{1.00}{(3.07\times10^{-4})^2} = -0.208\text{ V}$$

$$E_{\text{left}} = 0.000\text{ V}$$

$$E_{\text{cell}} = -0.208 - 0.000 = \underline{\underline{-0.208\text{ V}}}$$

Reduction on the left, oxidation on the right.

(f)

$$E_{\text{right}} = 0.771 - 0.0592\log\frac{0.1134}{0.003876} = 0.684\text{ V}$$

$$E_{\text{left}} = 0.334 - \frac{0.0592}{2}\log\frac{6.37\times10^{-2}}{7.93\times10^{-3}\,(1.16\times10^{-3})^4} = -0.040\text{ V}$$

$$E_{cell} = E_{right} - E_{left} = 0.684 - (-0.040) = \underline{\underline{0.724\text{ V}}}$$

Oxidation on the left, reduction on the right.

17-9. **(a)**

$$E_{Pb^{2+}} = -0.126 - \frac{0.0592}{2}\log\frac{1}{0.0848} = -0.158\text{ V}$$

$$E_{Zn^{2+}} = -0.763 - \frac{0.0592}{2}\log\frac{1}{0.1364} = -0.789\text{ V}$$

$$E_{cell} = E_{Pb^{2+}} - E_{Zn^{2+}} = -0.158 - (-0.789) = \underline{\underline{0.631\text{ V}}}$$

(c) $E_{SHE} = 0.000\text{ V}$

$$E_{TiO_2^{2+}} = 0.099 - 0.0592\log\frac{0.02723}{1.46\times10^{-3}\,(10^{-3})^2} = -0.331\text{ V}$$

$$E_{cell} = E_{SHE} - E_{TiO_2^{2+}} = 0.000 - (-0.331) = \underline{\underline{0.331\text{ V}}}$$

17-11. Note that in these calculations, it is necessary to round the answers to either one or two significant figures because the final step involves taking an antilogarithm of a large number (see page 73, text).

(a) $Fe^{3+} + V^{2+} = Fe^{2+} + V^{3+}$ $\quad E^0_{Fe^{3+}} = 0.771 \quad E_{V^{3+}} = -0.256$

$$0.771 - \frac{0.0592}{1}\log\frac{[Fe^{2+}]}{[Fe^{3+}]} = -0.256 - \frac{0.0592}{1}\log\frac{[V^{2+}]}{[V^{3+}]}$$

$$\frac{0.771-(-0.256)}{0.0592} = \log\frac{[V^{3+}][Fe^{2+}]}{[V^{2+}][Fe^{3+}]} = \log K_{eq} = 17.348$$

$$K_{eq} = 2.23\times10^{17} = \underline{\underline{2.2\times10^{17}}} = \frac{[Fe^{2+}][V^{3+}]}{[Fe^{3+}][V^{2+}]}$$

(c) $2V(OH)_4^+ + U^{4+} \rightleftarrows 2VO^{2+} + UO_2^{2+} + 4H_2O$ $\quad E^0_V = 1.00 \quad E^0_U = 0.334$

$$1.00 - \frac{0.0592}{2}\log\frac{[VO^{2+}]^2}{[V(OH)_4^+]^2[H^+]^4} = 0.334 - \frac{0.0592}{2}\log\frac{[U^{4+}]}{[UO_2^{2+}][H^+]^4}$$

$$\frac{2(1.00-0.334)}{0.0592} = \log\frac{[VO^{2+}]^2[UO_2^{2+}][H^+]^4}{[V(OH)_4^{2+}]^2[U^{4+}][H^+]^4} = \log K_{eq} = 22.50$$

$$K_{eq} = 3.2\times10^{22} = \underline{\underline{3\times10^{22}}} = \frac{[VO^{2+}]^2[UO_2^{2+}]}{[V(OH)_4^+]^2[U^{4+}]}$$

(e) $2Ce^{4+} + H_3AsO_3 + H_2O \rightleftarrows 2Ce^{3+} + H_3AsO_4 + 2H^+$

In 1 M $HClO_4$, $Ce^{4+} + e^- \rightleftarrows Ce^{3+}$ $\quad E^{0'} = 1.70$ V

$H_3AsO_4 + 2H^+ + 2e^- \rightleftarrows H_3AsO_3 + H_2O$ $\quad E^{0'} = 0.577$ V

$$1.70 - \frac{0.0592}{2}\log\frac{[Ce^{3+}]^2}{[Ce^{4+}]^2} = 0.577 - \frac{0.0592}{2}\log\frac{[H_3AsO_3]}{[H_3AsO_4][H^+]^2}$$

$$\frac{2(1.70-0.577)}{0.0592} = \log\frac{[Ce^{3+}]^2[H_3AsO_4]}{[Ce^{4+}]^2[H_3AsO_3][H^+]^2} = \log K_{eq}$$

$\log K_{eq} = 37.94$ and $K_{eq} = 8.9\times10^{37} = \underline{\underline{9\times10^{37}}}$

(g) $VO^{2+} + V^{2+} + 2H^+ = 2V^{3+} + H_2O$ $\quad E^0_{VO} = 0.359$ $\quad E^0_{V^{2+}} = -0.256$

$$0.359 - \frac{0.0592}{1}\log\frac{[V^{3+}]}{[VO^{2+}][H^+]^2} = -0.256 - \frac{0.0592}{1}\log\frac{[V^{2+}]}{[V^{3+}]}$$

$$\frac{0.359-(-0.256)}{0.0592} = \log\frac{[V^{3+}][V^{3+}]}{[VO^{2+}][H^+]^2[V^{2+}]} = \log K_{eq} = 10.389$$

$$K_{eq} = 2.446\times10^{10} = \underline{\underline{2.4\times10^{10}}} = \frac{[V^{3+}]^2}{[VO^{2+}][V^{2+}][H^+]^2}$$

17-12. At equivalence, $[Fe^{2+}] = [V^{3+}]$ $[Fe^{3+}] = [V^{2+}]$

(a)

$$E_{eq} = E^0_{Fe} - \frac{0.0592}{1}\log\frac{[Fe^{2+}]}{[Fe^{3+}]} = 0.771 - 0.0592\log\frac{[Fe^{2+}]}{[Fe^{3+}]}$$

$$E_{eq} = E^0_V - \frac{0.0592}{1}\log\frac{[V^{2+}]}{[V^{3+}]} = -0.256 - 0.0592\log\frac{[V^{2+}]}{[V^{3+}]}$$

$$2E_{eq} = E^0_{Fe} + E^0_V - 0.0592\log\frac{[Fe^{2+}][V^{2+}]}{[Fe^{3+}][V^{3+}]}$$

$$E_{eq} = \frac{0.771 - 0.256}{2} = \underline{\underline{0.258\text{ V}}}$$

(c) At equivalence, $[VO^{2+}] = 2[UO_2^{2+}]$ and $[V(OH)_4^+] = 2[U^{4+}]$

$$E_{eq} = 1.00 - \frac{0.0592}{1}\log\frac{[VO^{2+}]}{[V(OH)_4^+][H^+]^2}$$

$$2E_{eq} = 2\times 0.344 - 0.0592\log\frac{[U^{4+}]}{[UO_2^{2+}][H^+]^4}$$

$$3E_{eq} = 1.688 - 0.0592\log\frac{[VO^{2+}][U^{4+}]}{[V(OH)_4^+][UO_2^{2+}][H^+]^6}$$

$$= 1.688 - 0.0592\log\frac{2[UO_2^{2+}][U^{4+}]}{2[U^{4+}][UO_2^{2+}][H^+]^6}$$

$$3E_{eq} = 1.688 - 0.0592\log\frac{1}{(0.100)^6} = 1.688 - 0.355 = 1.333$$

$$E_{eq} = 1.333/3 = \underline{\underline{0.444\text{ V}}}$$

(e) At equivalence,

$$[Ce^{3+}] = 2[H_3AsO_4] \qquad [Ce^{4+}] = 2[H_3AsO_3] \qquad [H^+] = 1.00$$

$$E_{eq} = 1.70 - 0.0592\log\frac{[Ce^{3+}]}{[Ce^{4+}]}$$

$$2E_{eq} = 2\times 0.577 - 0.0592\log\frac{[H_3AsO_3]}{[H_3AsO_4][H^+]^2}$$

$$3E_{eq} = 2.854 - 0.0592\log\frac{[Ce^{3+}][H_3AsO_3]}{[Ce^{4+}][H_3AsO_4][H^+]^2}$$

$$E_{eq} = 0.951 - \frac{0.0592}{3}\log\frac{2[H_3AsO_4][H_3AsO_3]}{2[H_3AsO_3][H_3AsO_4](1.00)^2} = \underline{\underline{0.951\ \text{V}}}$$

(g) At equivalence, $[VO^{2+}] = [V^{2+}]$

$$E_{eq} = 0.359 - 0.0592\log\frac{[V^{3+}]}{[VO^{2+}][H^+]^2}$$

$$E_{eq} = -0.256 - 0.0592\log\frac{[V^{2+}]}{[V^{3+}]}$$

$$2E_{eq} = 0.103 - 0.0592\log\frac{[V^{3+}][V^{2+}]}{[VO^{2+}][V^{3+}][H^+]^2}$$

$$E_{eq} = 0.0515 - \frac{0.0592}{2}\log\frac{1}{1\times 10^{-2}} = \underline{\underline{-0.008\ \text{V}}}$$

17-14.

	E_{eq}, V	Indicator
(a)	0.258	Phenosafranine
(c)	0.444	Indigo tetrasulfonate or Methylene blue
(e)	0.951	Erioglaucin A
(g)	-0.008	None

17-15. (a) Two moles of V^{2+} react with one mole of Sn^{4+}.

Pre-equivalence point region

$$[V^{3+}] = \frac{\text{no. of mmol of } Sn^{4+} \text{ added} \times 2}{\text{total volume of solution in mL}}$$

$$[V^{2+}] = \frac{\text{no. of mmol } V^{2+} \text{ initially present} - \text{no. of mmol } Sn^{4+} \text{ added} \times 2}{\text{total solution volume in mL}}$$

$$E = E^0_{V^{3+}/V^{2+}} - 0.0592 \log \frac{[V^{2+}]}{[V^{3+}]}$$

The results are shown in the spreadsheet for part (a).

Equivalence point

$$E = E^0_{V^{3+}/V^{2+}} - 0.0592 \log \frac{[V^{2+}]}{[V^{3+}]}$$

$$2E = 2E^0_{Sn^{4+}/Sn^{2+}} - 0.0592 \log \frac{[Sn^{2+}]}{[Sn^{4+}]}$$

$$3E = \left(E^0_{V^{3+}/V^{2+}} + 2E^0_{Sn^{4+}/Sn^{2+}}\right) - 0.0592 \log \frac{[V^{2+}][Sn^{2+}]}{[V^{3+}][Sn^{4+}]}$$

But at equivalence $[V^{2+}] = 2[Sn^{4+}]$ and $[V^{[3+]}] = 2[Sn^{2+}]$

$$E_{eq} = \frac{\left(E^0_{V^{3+}/V^{2+}} + 2E^0_{Sn^{4+}/Sr^{2+}}\right)}{3} - \frac{0.0592}{3} \log 1.00 - \frac{\left(E^0_{V^{3+}/V^{2+}} + 2E^0_{Sn^{4+}/Sn^{2+}}\right)}{3}$$

This result is shown in the spreadsheet for part (a).

Post-equivalence point region

$$c_{Sn^{2+}} = \frac{\text{no. of mmol } V^{2+} \text{ initially present}/2}{\text{total solution volume in mL}} \approx [Sn^{2+}]$$

$$c_{Sn^{4+}} = \frac{\text{no. of mmol } Sn^{4+} \text{ added} - \text{no. of mmol } V^{2+} \text{ initially present}/2}{\text{total solution volume in mL}} \approx [Sn^{4+}]$$

$$E = E^0_{Sn^{4+}/Sn^{2+}} - \frac{0.0592}{2}\log\frac{[Sn^{2+}]}{[Sn^{4+}]}$$

The results are shown in the spreadsheet for this part.

(c) The data for these titrations, which are shown in the spreadsheets for these parts, are obtained in the same way as those for part (a).

(e)

Pre-equivalence point region

$$c_{UO_2^{2+}} = \frac{\text{no. of mmol of } KMnO_4 \text{ added} \times 5 \text{ mmol } UO_2^{2+}/2 \text{ mmol } KMnO_4}{\text{total solution volume in mL}} \approx [UO_2^{2+}]$$

$$c_{U^{4+}} = \frac{\text{no. of mmol of } U^{4+} \text{ initially present} - \text{no. of mmol of } KMnO_4 \text{ added} \times 5/2}{\text{total solution volume in mL}} \approx [U^{4+}]$$

$$E = E^0_{UO_2^{2+}/U^{4+}} - \frac{0.0592}{2}\log\frac{[U^{4+}]}{[UO_2^{2+}][H^+]^4}$$

The results are shown in the spreadsheet for part (e).

Equivalence point

$$E_{eq} = \frac{2E^0_{UO_2^{2+}/U^{4+}} + 5E^0_{MnO_4^-/Mn^{2+}}}{7} - \frac{0.0592}{7}\log\frac{1}{[H^+]^{12}}$$

See the spreadsheet for this part.

Post-equivalence point region

$$[MnO_4^-] = \frac{\text{no. of mmol } KMnO_4 \text{ added} - \text{no. of mmol } U^{4+} \text{ initially present} \times 2/5}{\text{total solution volume in mL}}$$

$$[Mn^{2+}] = \frac{\text{no. of mmol } U^{4+} \text{ initially present} \times 2/5}{\text{total solution volume in mL}}$$

$$E = E^0_{\mathrm{MnO_4^-/Mn^{2+}}} - \frac{0.0592}{5}\log\frac{[\mathrm{Mn^{2+}}]}{[\mathrm{MnO_4^-}][\mathrm{H^+}]^8}$$

The results are shown in the spreadsheet for part (e).

	A	B	C	D	E	F	G	H
1	**17-15 (a) Titration of 50.00 mL of 0.1000 M V^{2+} with 0.0500 M Sn^{4+}**							
2		Reaction: $2V^{2+} + Sn^{4+} \rightarrow 2V^{3+} + Sn^{2+}$						
3	For V^{3+}/V^{2+}, E^0	-0.256						
4	For Sn^{4+}/Sn^{2+}, E^0	0.154						
5	Initial conc. V^{2+}	0.1000						
6	Conc. Sn^{4+}	0.0500						
7	Volume solution, mL	50.00						
8	**Vol. Sn^{4+}, mL**	**$[V^{3+}]$**	**$[V^{2+}]$**	**$[Sn^{4+}]$**	**$[Sn^{2+}]$**	**E, V**		
9	10.00	0.0167	0.0667			-0.292		
10	25.00	0.0333	0.0333			-0.256		
11	49.00	0.0495	0.0010			-0.156		
12	49.90	0.0499	0.0001			-0.096		
13	50.00					0.017		
14	50.10			5.00E-05	0.0250	0.074		
15	51.00			4.95E-04	0.0248	0.104		
16	60.00			4.55E-03	0.0227	0.133		
17	**Documentation**							
18	Cell B9=$B$6*A9*2/($B$7+A9)							
19	Cell C9=($B$5*$B$7-$B$6*A9*2)/($B$7+A9)							
20	Cell F9=$B$3-0.0592*LOG10(C9/B9)							
21	Cell F13=($B$3+2*$B$4)/3							
22	Cell D14=($B$6*A14-$B$5*$B$7/2)/($B$7+A14)							
23	Cell E14=$B$7*$B$5/(2*($B$7+A14))							
24	Cell F14=$B$4-(0.0592/2)*LOG10(E14/D14)							

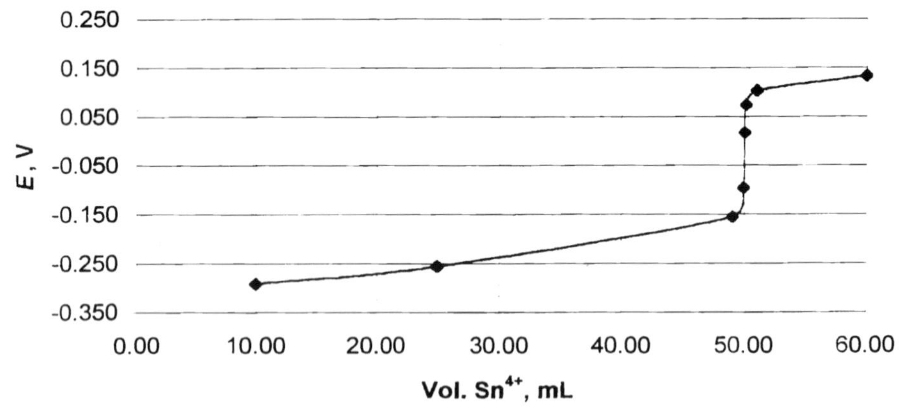

E, V
0.250
0.150
0.050
-0.050
-0.150
-0.250
-0.350
0.00 10.00 20.00 30.00 40.00 50.00 60.00
Vol. Sn^{4+}, mL

	A	B	C	D	E	F	G
1	**17-15 (c) Titration of 50.00 mL of 0.1000 M $Fe(CN)_6^{4-}$ with 0.0500 M Tl^{3+}**						
2		Reaction: $2Fe(CN)_6^{4-} + Tl^{3+} \rightarrow 2Fe(CN)_6^{3-} + Tl^{+}$					
3	For $Fe(CN)_6^{4-}$, E^0	0.36					
4	For Tl^{3+}/Tl^{+}, E^0	1.25					
5	Initial conc. $Fe(CN)_6^{4-}$	0.1000					
6	Conc. Tl^{3+}	0.0500					
7	Volume solution, mL	50.00					
8	**Vol. Tl^{3+}, mL**	**$[Fe(CN)_6^{4-}]$**	**$[Fe(CN)_6^{3-}]$**	**$[Tl^{3+}]$**	**$[Tl^{+}]$**	***E*, V**	
9	10.00	0.0667	0.0167			0.32	
10	25.00	0.0333	0.0333			0.36	
11	49.00	0.0010	0.0495			0.46	
12	49.90	0.0001	0.0499			0.52	
13	50.00					0.95	
14	50.10			5.00E-05	0.0250	1.17	
15	51.00			4.95E-04	0.0248	1.20	
16	60.00			4.55E-03	0.0227	1.23	
17	**Documentation**						
18	Cell B9=($B$5*$B$7-$B$6*A9*2)/($B$7+A9)						
19	Cell C9=$B$6*A9*2/($B$7+A9)						
20	Cell F9=$B$3-0.0592*LOG10(B9/C9)						
21	Cell F13=($B$3+2*$B$4)/3						
22	Cell D14=($B$6*A14-$B$5*$B$7/2)/($B$7+A14)						
23	Cell E14=($B$5*$B$7/2)/($B$7+A14)						
24	Cell F14=$B$4-(0.0592/2)*LOG10(E14/D14)						

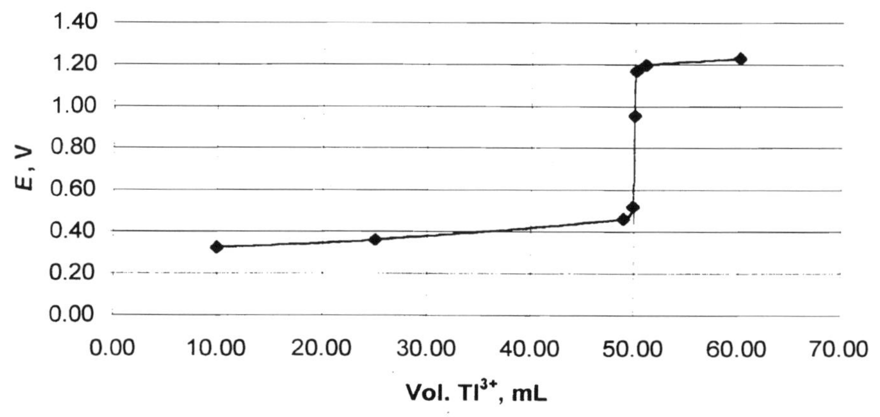

	A	B	C	D	E	F	G
1	**17-15 (e) Titration of 50.00 mL 0.05000 M U^{4+} with 0.02000 M MnO_4^-**						
2		Reaction: $2MnO_4^- + 5U^{4+} + 2H_2O \rightarrow 2Mn^{2+} + 5UO_2^{2+} + 4H^+$					
3	For U^{4+}/UO_2^{2+}, E^0	0.334					
4	For MnO_4^-, E^0	1.51					
5	Initial conc. U^{4+}	0.0500					
6	Conc. MnO_4^-	0.0200					
7	Volume solution, mL	50.00					
8	**Vol. MnO_4^-, mL**	**$[U^{4+}]$**	**$[UO_2^{2+}]$**	**$[MnO_4^-]$**	**$[Mn^{2+}]$**	**$[H^+]$**	***E*, V**
9	10.00	0.0333	0.0083			1.00	0.316
10	25.00	0.0167	0.0167			1.00	0.334
11	49.00	0.0005	0.0247			1.00	0.384
12	49.90	0.0001	0.0250			1.00	0.414
13	50.00					1.00	1.17
14	50.10			2.00E-05	0.0100	1.00	1.48
15	51.00			0.0002	0.0099	1.00	1.49
16	60.00			0.0018	0.0091	1.00	1.50
17	**Documentation**						
18	Cell B9=$B$5*$B$7-$B$6*A9*5/2)/($B$7+A9)						
19	Cell C9=($B$6*A9*5/2)/($B$7+A9)						
20	Cell F9=1.00 (entry)						
21	Cell G9=$B$3-(0.0592/2)*LOG10(B9/(C9*F9^4))						
22	Cell G13=((5*$B$4+2*$B$3)/7)-(0.0592/7)*LOG10(F13)						
23	Cell D14=($B$6*A14-$B$5*$B$7*2/5)/($B$7+A14)						
24	Cell E14=($B$5*$B$7*2/5)/($B$7+A14)						
25	Cell G14=$B$4-(0.0592/5)*LOG10(E14/(D14*F14^8))						

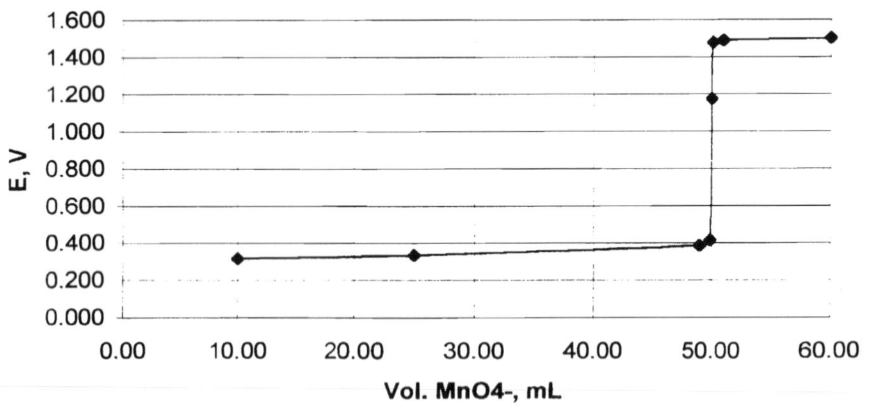

E, V: 0.000, 0.200, 0.400, 0.600, 0.800, 1.000, 1.200, 1.400, 1.600

Vol. MnO4-, mL: 0.00, 10.00, 20.00, 30.00, 40.00, 50.00, 60.00

Chapter 18

18-1. **(a)** $2Mn^{2+} + 5S_2O_8^{2-} + 8H_2O \rightarrow 10SO_4^{2-} + 2MnO_4^- + 16H^+$

(b) $NaBiO_3(s) + 2Ce^{3+} + 4H^+ \rightarrow BiO^+ + 2Ce^{4+} + 2H_2O + Na^+$

(c) $H_2O_2 + U^{4+} \rightarrow UO_2^{2+} + 2H^+$

(d) $V(OH)_4^+ + Ag(s) + Cl^- + 2H^+ \rightarrow VO^{2+} + AgCl(s) + 3H_2O$

(e) $2MnO_4^- + 5H_2O_2 + 6H^+ \rightarrow 5O_2 + 2Mn^{2+} + 8H_2O$

(f) $ClO_3^- + 6I^- + 6H^+ \rightarrow 3I_2 + Cl^- + 3H_2O$

18-3. Only in the presence of Cl^- ion is Ag a sufficiently good reducing agent to be very useful for prereductions. In the presence of Cl^- ion, the half-reaction occurring in the Walden reductor is

$$Ag(s) + Cl^- \rightarrow AgCl(s) + e^-$$

The excess HCl increases the tendency of this reaction to occur by the common ion effect.

18-5. $UO_2^{2+} + 2Ag(s) + 4H^+ + 2Cl^- \rightleftarrows U^{4+} + 2AgCl(s) + H_2O$

18-7. Standard solutions of reductants find somewhat limited use because of their susceptibility to air oxidation.

18-8. Standard $KMnO_4$ solutions are seldom used to titrate solutions containing HCl because of the tendency of MnO_4^- to oxidize Cl^- to Cl_2, thus causing an over-consumption of MnO_4^-.

18-10. $2MnO_4^- + 3Mn^{2+} + 2H_2O \rightarrow 5MnO_2(s) + 4H^+$

18-13. $4MnO_4^- + 2H_2O \rightarrow \underset{\text{brown}}{4MnO_2(s)} + 3O_2 + 4OH^-$

18-15. Iodine is not sufficiently soluble in water to produce a useful standard reagent. It is quite soluble in solutions that contain an excess of iodide, however, as a consequence of the formation of the triiodide complex. The rate at which iodine dissolves in iodide solutions increases as the concentration of iodide ion becomes greater. For this reason, iodine is always dissolved in a very concentrated solution of potassium iodide and diluted only after solution is complete.

18-17. $S_2O_3^{2-} + H^+ \rightarrow HSO_3^- + S(s)$

18-19. $BrO_3^- + \underset{\text{excess}}{6I^-} + 6H^+ \rightarrow Br^- + 3I_2 + 3H_2O$

$I_2 + 2S_2O_3^{2-} \rightarrow 2I^- + S_4O_6^{2-}$

18-21. $2I_2 + N_2H_4 \rightarrow N_2 + 4H^+ + 4I^-$

18-23.

$$0.2464\text{ g} \times \frac{1\text{ mmol Fe}}{0.055847\text{ g Fe}} = 4.4121\text{ mmol Fe}$$

(a) $$\frac{4.4121\text{ mmol Fe}}{39.31\text{ mL}} \times \frac{\text{mmol Ce}^{4+}}{\text{mmol Fe}} = \underline{\underline{0.1122\text{ M Ce}^{4+}}}$$

(c) $$\frac{4.4121\text{ mmol Fe}}{39.31\text{ mL}} \times \frac{\text{mmol MnO}_4^-}{5\text{ mmol Fe}} = \underline{\underline{0.02245\text{ M MnO}_4^-}}$$

(e) $$\frac{4.4121\text{ mmol Fe}}{39.31\text{ mL}} \times \frac{\text{mmol IO}_3^-}{4\text{ mmol Fe}} = \underline{\underline{0.02806\text{ M IO}_3^-}}$$

18-24.

$$250\text{ mL} \times 0.03500\frac{\text{mmol K}_2\text{Cr}_2\text{O}_7}{\text{mL}} \times \frac{0.294185\text{ g}}{\text{mmol K}_2\text{Cr}_2\text{O}_7} = 2.57412\text{ g}$$

Dissolve 2.574 g $K_2Cr_2O_7$ in sufficient water to give 250.0 mL of solution.

18-26.

$$1.5\text{ L} \times \frac{0.1\text{ mol KMnO}_4}{\text{L}} \times \frac{158.034\text{ g}}{\text{mol KMnO}_4} = 23.705\text{ g}$$

Dissolve about 24 g of $KMnO_4$ in 1.5 L of water.

18-28. $$\frac{0.1467\text{ g Na}_2\text{Ox}}{28.85\text{ mL KMnO}_4}\times\frac{1\text{ mmol Na}_2\text{Ox}}{0.133999\text{ g Na}_2\text{Ox}}\times\frac{2\text{ mmol KMnO}_4}{5\text{ mmol Na}_2\text{Ox}} = \underline{\underline{0.01518\text{ M KMnO}_4}}$$

18-30. $$Cr_2O_7^{2-} + 6I^- + 14H^+ = 2Cr^{3+} + 3I_2 + 7H_2O$$

$$1\text{ mmol Cr}_2\text{O}_7^{2-} \equiv 3\text{ mmol I}_2 \equiv 6\text{ mmol S}_2\text{O}_3^{2-}$$

$$\frac{0.1518\text{ g K}_2\text{Cr}_2\text{O}_7}{46.13\text{ mL Na}_2\text{S}_2\text{O}_3}\times\frac{1\text{ mmol K}_2\text{Cr}_2\text{O}_7}{0.294185\text{ g K}_2\text{Cr}_2\text{O}_7}\times\frac{6\text{ mmol S}_2\text{O}_3^{2-}}{\text{mmol K}_2\text{Cr}_2\text{O}_7} = \underline{\underline{0.06711\text{ M Na}_2\text{S}_2\text{O}_3}}$$

18-32. $$1\text{ mmol I}_2 \equiv 1\text{ mmol Sb} \equiv 2\text{mmol Sb}_2\text{S}_3$$

$$\frac{41.67\text{ mL}\times 0.03134\frac{\text{mmol I}_2}{\text{mL I}_2}\times\frac{1\text{ mmol Sb}}{\text{mmol I}_2}\times 0.12176\frac{\text{g Sb}}{\text{mmol Sb}}}{1.080\text{ g sample}}\times 100\,\% = \underline{\underline{14.72\%\text{ Sb}}}$$

$$\frac{41.67\times 0.03134\times\frac{1}{2}\times 0.33971}{1.080}\times 100\,\% = \underline{\underline{20.54\%\text{ Sb}_2\text{S}_3}}$$

18-34. Letting $A = CS(NH_2)_2$, $4\text{ mol KBrO}_3 = 3\text{ mol A}$

$$\frac{14.1\text{ mL KBrO}_3}{0.0715\text{ g sample}}\times 0.00833\times\frac{\text{mmol KBrO}_3}{\text{mL KBrO}_3}\times\frac{3\text{ mmol A}}{4\text{ mmol KBrO}_3}\times$$

$$\frac{0.076122\text{ g A}}{\text{mmol A}}\times 100\,\% = \underline{\underline{9.38\%\text{ A}}}$$

18-35. $$1\text{ mmol KMnO}_4 \equiv 5\text{ mmol Fe} \equiv \frac{5}{2}\text{mmol Fe}_2\text{O}_3$$

$$\text{no. mmol KMnO}_4 = 39.21\text{ mL KMnO}_4\times\frac{0.02086\text{ mmol KMnO}_4}{\text{mL KMnO}_4} = 0.81792$$

(a) $$\frac{0.81792\text{ mmol KMnO}_4}{0.7120\text{ g sample}}\times\frac{5\text{ mmol Fe}}{\text{mmol KMnO}_4}\times\frac{0.055847\text{ g Fe}}{\text{mmol Fe}}\times 100\,\% = \underline{\underline{32.08\%\text{ Fe}}}$$

(b) $$\frac{0.81792\text{ mmol KMnO}_4}{0.7120\text{ g sample}}\times\frac{5\text{ mmol Fe}_2\text{O}_3}{2\text{ mmol KMnO}_4}\times\frac{0.159692\text{ g Fe}_2\text{O}_3}{\text{mmol Fe}_2\text{O}_3}\times 100\,\% \;=\; \underline{\underline{45.86\%\text{ Fe}_2\text{O}_3}}$$

18-37. $1\text{ mmol K}_2\text{Cr}_2\text{O}_7 \;\equiv\; 6\text{ mmol Fe} \;\equiv\; 3\text{ mmol H}_2\text{NOH}$

$$\frac{23.61\text{ mL K}_2\text{Cr}_2\text{O}_7}{50.00\text{ mL sample}}\times 0.02170\frac{\text{mmol K}_2\text{Cr}_2\text{O}_7}{\text{mL K}_2\text{Cr}_2\text{O}_7}\times\frac{3\text{ mmol H}_2\text{NOH}}{\text{mmol K}_2\text{Cr}_2\text{O}_7} \;=\; \underline{\underline{0.03074\text{ M}}}$$

18-39.

$$\text{no. mmol Fe}^{2+} \;=\; 50.00\text{ mL Fe}^{2+}\times 0.09601\frac{\text{mmol Fe}^{2+}}{\text{mmol Fe}^{2+}} \;=\; 4.8005$$

$$\text{no. mmol Fe}^{2+}\text{ titrated by Ce}^{4+} \;=\; 12.99\text{ mL Ce}^{4+}\times 0.08362\frac{\text{mmol Ce}^{4+}}{\text{mL}}\times\frac{1\text{ mmol Fe}^{2+}}{\text{mmol Ce}^{4+}}$$

$$=\; 1.08622$$

$$\text{no. mmol Fe}^{2+}\text{ consumed by analyte} \;=\; 4.8005 - 1.08622 \;=\; 3.71428$$

$$\frac{3.71428\text{ mmol Fe}^{2+}}{0.1342\text{ g sample}}\times\frac{1\text{ mmol KClO}_3}{6\text{ mmol Fe}^{2+}}\times\frac{0.122549\text{ g KClO}_3}{\text{mmol KClO}_3}\times 100\,\% \;=\; \underline{\underline{56.53\%\text{ KClO}_3}}$$

18-41. $\text{H}_3\text{AsO}_3 + \text{I}_2 + \text{H}_2\text{O} \;\rightleftarrows\; \text{H}_3\text{AsO}_4 + 2\text{I}^- + 2\text{H}^+$

$2\text{ mmol I}_2 \;\equiv\; 2\text{ mmol H}_2\text{AsO}_3 \;\equiv\; 1\text{ mmol As}_2\text{O}_3$

$$\frac{23.77\text{ mL I}_2}{8.13\text{ g sample}}\times 0.02425\frac{\text{mmol I}_2}{\text{mL I}_2}\times\frac{1\text{ mmol As}_2\text{O}_3}{2\text{ mmol I}_2}\times\frac{0.197841\text{ g As}_2\text{O}_3}{\text{mmol As}_2\text{O}_3}\times 100\,\% \;=\; \underline{\underline{0.701\%\text{ As}_2\text{O}_3}}$$

18-43.

$$\text{no. mmol I}_2\text{ taken} \;=\; 50.0\text{ mL}\times 0.01194\frac{\text{mmol I}_2}{\text{mL I}_2} \;=\; 0.5970$$

$$\text{no. mmol I}_2 \text{ consumed by Na}_2\text{S}_2\text{O}_3 = 16.77 \text{ mL} \times 0.01325 \frac{\text{mmol}}{\text{mL}} \times \frac{1 \text{ mmol I}_2}{2 \text{ mmol Na}_2\text{S}_2\text{O}_3}$$

$$= 0.1111$$

$$\text{no. mmol I}_2 \text{ consumed by analyte} = 0.5970 - 0.1111 = 0.4859$$

$$\frac{0.4859 \text{ mmol I}_2}{1.657 \text{ g sample}} \times \frac{2 \text{ mmol C}_2\text{H}_5\text{SH}}{\text{mmol I}_2} \times 0.06213 \frac{\text{g C}_2\text{H}_5\text{SH}}{\text{mmol C}_2\text{H}_5\text{SH}} \times 100\% = \underline{\underline{3.64\% \text{ C}_2\text{H}_5\text{SH}}}$$

18-45. $1 \text{ mmol KI} \equiv 1 \text{ mmol IO}_3^- \equiv 3 \text{ mmol I}_2 \equiv 6 \text{ mmol Na}_2\text{S}_2\text{O}_3$

$$\frac{20.66 \text{ mL Na}_2\text{S}_2\text{O}_3}{1.204 \text{ g sample}} \times 0.05551 \frac{\text{mmol Na}_2\text{S}_2\text{O}_3}{\text{mL}} \times \frac{\text{mmol KI}}{6 \text{ mmol Na}_2\text{S}_2\text{O}_3} \times$$

$$0.16600 \frac{\text{g KI}}{\text{mmol KI}} \times 100\% = \underline{\underline{2.635\% \text{ KI}}}$$

18-46.

$$\text{no. mmol Fe} = (13.72 \times 0.01920) \text{ mmol KMnO}_4 \times \frac{5 \text{ mmol Fe}}{\text{mmol KMnO}_4} \times \frac{500.0 \text{ mL}}{50.00 \text{ mL}}$$

$$= 13.171 \text{ mmol Fe}$$

Similarly,

$$\text{no. mmol (Fe + Cr)} = 36.43 \times 0.01920 \times 5 \times 500/100 = 17.486$$

$$\text{no. mmol Cr} = 17.486 - 13.171 = 4.315$$

$$\frac{13.171 \text{ mmol Fe} \times 0.055847 \text{ g Fe/mmol}}{1.065 \text{ g sample}} \times 100\% = \underline{\underline{69.07\% \text{ Fe}}}$$

$$\frac{4.315 \text{ mmol Cr} \times 0.051996 \text{ g Cr/mmol}}{1.065 \text{ g sample}} \times 100\% = \underline{\underline{21.07\% \text{ Cr}}}$$

18-48. $6 \text{ mmol Fe}^{2+} \equiv 1 \text{ mmol Cr}_2\text{O}_7^{2-} \equiv 4 \text{ mmol Tl}^+$

$$(40.60 \times 0.1004) \text{ mmol Fe} \times \frac{4 \text{ mmol Tl}}{6 \text{ mmol Fe}} \times 0.20438 \frac{\text{g Tl}}{\text{mmol Tl}} = \underline{\underline{0.5554 \text{ g Tl}}}$$

18-49. $1 \text{ mmol } IO_3^- \equiv 2 \text{ mmol } H_2SO_3 \equiv 2 \text{ mmol } SO_2$

In $2.50 \frac{\text{L}}{\text{min}} \times 64.0 \text{ min} = 160.0 \text{ L}$ of sample, there are

$$4.98 \text{ mL } IO_3^- \times 0.003125 \frac{\text{mmol } IO_3^-}{\text{mL } IO_3^-} \times \frac{2 \text{ mmol } SO_2}{\text{mmol } IO_3^-} \times \frac{0.064065 \text{ g } SO_2}{\text{mmol } SO_2} = 0.0019940 \text{ g } SO_2$$

$$\frac{0.001994 \text{ g } SO_2}{160.0 \text{ L} \times 1.2 \text{ g/L}} \times 10^6 \text{ ppm} = 10.39 \cong \underline{\underline{10.4 \text{ ppm } SO_2}}$$

18-51. $1 \text{ mmol } I_2 \equiv 1 \text{ mmol } H_2S \equiv 2 \text{ mmol } Na_2S_2O_3$

$$\text{no. mmol } I_2 \text{ taken} = 10.00 \text{ mL} \times 0.01070 \frac{\text{mmol } I_2}{\text{mL}} = 0.1070$$

$$\text{no. mmol } I_2 \text{ in excess} = 12.85 \text{ mL} \times 0.01344 \frac{\text{mmol } Na_2S_2O_3}{\text{mL}} \times \frac{1 \text{ mmol } I_2}{2 \text{ mmol } Na_2S_2O_3} = 0.08635$$

$$\text{no. mmol } I_2 = \text{mmol } H_2S = 0.1070 - 0.08635 = 0.02065$$

$$\frac{0.02065 \text{ mmol } H_2S}{30.00 \text{ L sample} \times 1.2 \text{ g sample/L}} \times 0.034082 \frac{\text{g } H_2S}{\text{mmol}} \times 10^6 \text{ ppm} = \underline{\underline{19.5 \text{ ppm } H_2S}}$$

18-53. $O_2 + 4Mn(OH)_2(s) + 2H_2O \rightarrow 4\,Mn(OH)_3(s)$

$$4\,Mn(OH)_3(s) + 12H^+ + 4I^- \rightarrow 4Mn^{2+} + 2I_2 + 6\,H_2O$$

$$1 \text{ mmol } O_2 \equiv 2 \text{ mmol } I_2 \equiv 4 \text{ mmol} S_2O_3^{2-}$$

$$12.7 \text{ mL} \times 0.00962 \text{ mmol } S_2O_3^{2-} \times \frac{1 \text{ mmol } O_2}{4 \text{ mmol } S_2O_3^{2-}} \times \frac{32.0 \text{ mg } O_2}{\text{mmol } O_2} = 9.774 \text{ mg } O_2$$

$$\frac{9.774 \text{ mg } O_2}{254 \text{ mL sample} \times 25.0/254} = \underline{\underline{0.0397 \text{ mg } O_2/\text{mL sample}}}$$

Chapter 19

19-1. **(a)** An indicator electrode is an electrode used in potentiometry that responds to variations in the activity of an analyte ion or molecule.

(c) An electrode of the first kind is a metal electrode that is used to determine the concentration of its cation in a solution.

19-2. **(a)** A liquid-junction potential is the potential that develops across the interface between two solutions having different electrolyte compositions.

19-3. **(a)** An electrode of the first kind for Hg(II) would take the form

$\|Hg^{2+}(x\text{ M})|Hg$

$$E_{Hg} = E^0_{Hg} - \frac{0.0592}{2}\log\frac{1}{[Hg^{2+}]} = E^0_{Hg} + \frac{0.0592}{2}\text{pHg}$$

(b) An electrode of the second kind for EDTA would take the form

$\|HgY^{2-}(y\text{ M}),Y^{4-}(x\text{ M})|Hg$

where a small and fixed amount of HgY^{2-} is introduced into the analyte solution. Here the potential of the mercury electrode is given by

$$E_{Hg} = K - \frac{0.0592}{2}\log[Y^{4-}] = K + \frac{0.0592}{2}\text{pY}$$

where

$$K = E^0_{HgY^{2-}} - \frac{0.0592}{2}\log\frac{1}{a_{HgY^{2-}}} \cong 0.21 - \frac{0.0592}{2}\log\frac{1}{[HgY^{2-}]}$$

19-5. The pH-dependent potential that develops across a glass membrane arises from the difference in positions of dissociation equilibria that arise on each of the two surfaces. These equilibria are described by the equation

$$\underset{\text{membrane}}{H^+Gl^-} \rightleftharpoons \underset{\text{soln}}{H^+} + \underset{\text{membrane}}{Gl^-}$$

The surface exposed to the solution having the higher hydrogen ion concentration then becomes positive with respect to the other surface. This charge difference, or potential, serves as the analytical parameter when the pH of the solution on one side of the membrane is held constant.

19-7. Uncertainties that may be encountered in pH measurements include: (1) the acid error in highly acidic solutions, (2) the alkaline error in strongly basic solutions, (3) the error that arises when the ionic strength of the calibration standards differ from that of the analyte solution, (4) uncertainties in the pH of the standard buffers, (5) nonreproducible junction potentials when samples of low ionic strength are measured, and (6) dehydration of the working surface.

19-9. The alkaline error arises when a glass electrode is employed to measure the pH of solutions having pH values in the 10 to 12 range or greater. In the presence of alkali ions, the glass surface becomes responsive to not only hydrogen but alkali ions as well. Low pH values arise as a consequence.

19-11. (b) The *boundary potential* for a membrane electrode is a potential that develops when the membrane separates two solutions that have different concentrations of a cation or an anion that the membrane binds selectively. For an aqueous solution, the following equilibria develop when the membrane is positioned between two solutions of A^+:

$$\underset{\text{membrane}_1}{A^+M^-} \rightleftarrows \underset{\text{soln}_1}{A^+} + \underset{\text{membrance}_1}{M^-}$$

$$\underset{\text{membrane}_2}{A^+M^-} \rightleftarrows \underset{\text{soln}_2}{A^+} + \underset{\text{membrane}_2}{M^-}$$

where the subscripts refer to the two sides of the membrane. A potential develops across this membrane if one of these equilibria proceeds further to the right than the other, and this potential is the boundary potential. For example if the concentration of A^+ is greater in solution 1 than in solution 2, the negative charge on side 1 of the membrane will be less than that of side 2 because the equilibrium on side 1 will lie further to the left. Thus, a greater fraction of the negative charge on side 1 will be neutralized by A^+.

(d) The membrane in a solid-state electrode for F^- is crystalline LaF_3, which when immersed in aqueous solution LaF_3 dissociates according to the equation

$$LaF_3 \rightleftarrows La^{3+} + 3F^-$$

Thus, a boundary potential develops across this membrane when it separates two solutions of different F^- ion concentration. The source of this potential is described in part (b) of this answer.

19-12. The direct potentiometric measurement of pH provides a measure of the equilibrium concentration of hydronium ions present in a solution of the sample. A potentiometric titration provides information on the amount of reactive protons, both ionized and nonionized, that are present in a sample.

19-15. (a)

$$E_{Ag} = 0.799 - 0.0592 \log \frac{1}{[Ag^+]} \qquad K_{sp} = [Ag^+][IO_3^-] = 3.1 \times 10^{-8}$$

$$E = 0.799 - 0.0592 \log \frac{1}{K_{sp}} - 0.0592 \log [IO_3^-]$$

When $[IO_3^-] = 1.00$, E is equal to $E^0_{AgIO_3}$ for the reduction of $AgIO_3$, that is,

$$E^0_{AgIO_3} = 0.799 - 0.0592 \log \frac{1}{3.1 \times 10^{-8}} = \underline{\underline{0.354 \text{ V}}}$$

(b) SCE || IO_3^- (x M), $AgIO_3$(sat'd) | Ag

(c)

$$E_{cell} = E_{AgIO_3} - E_{SCE} = 0.354 - 0.0592 \log [IO_3^-] - 0.244$$

$$= 0.110 + 0.0592 \text{ pIO}_3$$

$$\underline{\underline{\text{pIO}_3 = \frac{E_{cell} - 0.110}{0.0592}}}$$

(d)

$$\text{pIO}_3 = \frac{0.294 - 0.110}{0.0592} = \underline{\underline{3.11}}$$

19-17. (a) SCE || SCN^- (x M), AgSCN(sat'd) | Ag

(c) SCE || SO_3^{2-} (x M), Ag_2SO_3(sat'd) | Ag

19-19. **(a)** $$\text{pSCN} = \frac{0.122 + 0.153}{0.0592} = \underline{\underline{4.65}}$$

(c) $$\text{pSO}_3 = \frac{2(0.300 - 0.146)}{0.0592} = \underline{\underline{5.20}}$$

19-20. $$Ag_2CrO_4(s) + 2e^- \rightleftarrows 2Ag(s) + CrO_4^{2-} \qquad E^0 = 0.446\text{ V}$$

$$0.402 = 0.446 - 0.0296\log[CrO_4^{2-}] - 0.244 = 0.202 + 0.0296\text{ pCrO}_4$$

$$\text{pCrO}_4 = (0.402 - 0.202)/0.0296 = \underline{\underline{6.76}}$$

19-21. Substituting into Equation 19-20 gives

$$\text{pH} = -\frac{E_{\text{cell}} - K}{0.0592/1} \quad \text{and} \quad 4.006 = -\frac{0.2094 - K}{0.0592}$$

$$K = 4.006 \times 0.0592 + 0.2094 = 0.44656$$

(a) $$\text{pH} = -(-0.3011 - 0.44656)/0.0592 = \underline{\underline{12.629}}$$

$$a_{H^+} = \text{antilog}(-12.629) = \underline{\underline{2.35 \times 10^{-13}}}$$

(b) $$\text{pH} = -(0.1163 - 0.44656)/0.0592 = \underline{\underline{5.579}}$$

$$a_{H^+} = \text{antilog}(-5.579) = \underline{\underline{2.64 \times 10^{-6}}}$$

(c) For part (a),

If $E = -0.3011 + 0.002 = -0.2991\text{ V}$

$$\text{pH} = -(-0.2991 - 0.44656)/0.0592 = \underline{\underline{12.596}}$$

$$a_{H^+} = \underline{\underline{2.54 \times 10^{-13}}}$$

$$\text{If } E = -0.3011 - 0.002 = -0.3031 \text{ V}$$

$$\text{pH} = -(-0.3031 - 0.44656)/0.0592 = \underline{\underline{12.663}}$$

$$a_{H^+} = \underline{\underline{2.17 \times 10^{-13}}}$$

Proceeding in the same way for part (b) we obtain

$$\text{pH} = \underline{\underline{5.545}} \text{ and } \underline{\underline{5.612}}$$

$$a_{H^+} = \underline{\underline{2.85 \times 10^{-6}}} \text{ and } \underline{\underline{2.44 \times 10^{-6}}}$$

19-22.

$$\text{no. mmol HA} = 27.22 \text{ mL NaOH} \times 0.1025 \frac{\text{mmol NaOH}}{\text{mL NaOH}} \times \frac{1 \text{ mmol HA}}{\text{mmol NaOH}} = 2.7901$$

$$\frac{0.3798 \text{ g HA}}{2.7901 \text{ mmol HA}} \times 10^3 \frac{\text{mmol}}{\text{mol}} = \underline{\underline{136 \frac{\text{g HA}}{\text{mol}}}} = \mathcal{M}_{HA}$$

19-24. For all, $E_{cell} = E_{ind} - E_{SCE}$

Note there are 2 mmol of Ce(IV) that react per mmol of HNO_2.

Pre-equivalence point region

$$[NO_3^-] \approx c_{NO_3^-} = \frac{\text{no. of mmol Ce(IV) added} \times \frac{1 \text{ mmol HNO}_2}{2 \text{ mmol Ce(IV)}}}{\text{total solution volume in mL}}$$

$$[HNO_2] \approx c_{HNO_2} = \frac{\text{mmol HNO}_2 \text{ initially present} - \text{mmol Ce(IV) added}/2}{\text{total solution volume in mL}}$$

$$E_{ind} = E^0_{NO_3^-/HNO_2} - \frac{0.0592}{2} \log \frac{[HNO_2]}{[NO_3^-][H^+]^3}$$

Equivalence point

$$E_{ind} = \left(\frac{2E_{NO_3^-/HNO_2} + E^0_{Ce^{4+}/Ce^{3+}}}{3} \right) - \frac{0.0592}{3} \log \left(\frac{1}{[H^+]^3} \right)$$

Post-equivalence point region

$$[\text{Ce(III)}] \approx c_{\text{Ce(IV)}} = \frac{\text{no. mmol HNO}_2\text{ originally present} \times \frac{2\text{ mmol Ce(III)}}{1\text{ mmol HNO}_2}}{\text{total solution volume in mL}}$$

$$[\text{Ce(IV)}] \approx c_{\text{Ce(IV)}} = \frac{\text{no. mmol Ce(IV)} - \text{no. mmol HNO}_2 \times \frac{2\text{ mmol Ce(IV)}}{1\text{ mmol HNO}_2}}{\text{total solution volume in mL}}$$

$$E_{\text{ind}} = E^0_{\text{Ce}^{4+}/\text{Ce}^{3+}} - 0.0592 \log \frac{[\text{Ce(III)}]}{[\text{Ce(IV)}]}$$

	A	B	C	D	E	F	G	H	I	J	K	O
1	**19-24 Titration of HNO_2 with Ce(IV) at pH 1.00**											
2	E^0 HNO_2	0.94	Reaction:									
3	E^0 Ce(IV)/Ce(III)	1.44	$HNO_2 + 2Ce(IV) + H_2O \rightarrow NO_3^- + 3H^+ + 2Ce(III)$									
4	Initial conc. HNO_2	0.0500	Equivalence point is at 0.0500 × 40.00 × 2/0.0800 = 50.00									
5	Conc. Ce(IV)	0.0800	mL of Ce(IV)									
6	Vol HNO_2	40.00										
7	Initial soln. vol.	75.00										
8	pH	1.00										
9	E_{SCE}	0.244							First Derivative			Second
10	Vol Ce(IV), mL	$[NO_3^-]$	$[HNO_2]$	[Ce(III)]	[Ce(IV)]	E_{ind}	E vs SCE	Midpoint Vol.	ΔE	ΔVol	$\Delta E/\Delta V$	$\Delta^2 E/\Delta V^2$
11	5.00	0.00250	0.02250			0.823	0.58					
12	10.00	0.00471	0.01882			0.833	0.59	7.50	0.010	5.00	0.002	
13	15.00	0.00667	0.01556			0.840	0.60	12.50	0.007	5.00	0.001	0.000
14	25.00	0.01000	0.01000			0.851	0.61	20.00	0.011	10.00	0.001	0.000
15	40.00	0.01391	0.00348			0.869	0.63	32.50	0.018	15.00	0.001	0.000
16	49.00	0.01581	0.00032			0.901	0.66	44.50	0.032	9.00	0.004	0.000
17	49.50	0.01590	0.00016			0.910	0.67	49.25	0.009	0.50	0.018	0.003
18	49.60	0.01592	0.00013			0.913	0.67	49.55	0.003	0.10	0.029	0.036
19	49.70	0.01594	0.00010			0.917	0.67	49.65	0.004	0.10	0.037	0.083
20	49.80	0.01596	6.41E-05			0.922	0.68	49.75	0.005	0.10	0.052	0.151
21	49.90	0.01598	3.20E-05			0.931	0.69	49.85	0.009	0.10	0.089	0.370
22	49.95	0.01599	1.60E-05			0.940	0.70	49.93	0.009	0.05	0.178	1.188
23	49.99	0.01600	3.20E-06			0.961	0.72	49.97	0.021	0.04	0.517	7.534
24	50.00					1.05	0.80	50.00	0.087	0.01	8.678	326.419
25	50.01			0.03200	6.40E-06	1.22	0.98	50.01	0.174	0.01	17.355	867.746
26	50.05			0.03199	3.20E-05	1.26	1.02	50.03	0.041	0.04	1.034	-652.838
27	50.10			0.03197	6.39E-05	1.28	1.04	50.08	0.018	0.05	0.356	-15.068
28	50.20			0.03195	1.28E-04	1.30	1.05	50.15	0.018	0.10	0.178	-2.376
29	50.30			0.03192	1.92E-04	1.31	1.06	50.25	0.010	0.10	0.104	-0.740
30	50.40			0.03190	2.55E-04	1.32	1.07	50.35	0.007	0.10	0.074	-0.303
31	50.50			0.03187	3.19E-04	1.32	1.08	50.45	0.006	0.10	0.057	-0.166
32	51.00			0.03175	6.35E-04	1.34	1.10	50.75	0.018	0.50	0.036	-0.072
33	60.00			0.02963	0.00593	1.40	1.15	55.50	0.059	9.00	0.007	-0.006
34	75.00			0.02667	0.01333	1.42	1.18	67.50	0.024	15.00	0.002	0.000
35	90.00			0.02424	0.01939	1.43	1.19	82.50	0.012	15.00	0.001	0.000
36												
37												

19-24.

	A	B	C	D	E	F	G	H	I	J	K	O
38	**Spreadsheet Documentation**											
39	Cell B11=($B$5*A11/2)/($B$7+A11)					Cell H12=(A12+A11)/2						
40	Cell C11=($B$4*$B$6-$B$5*A11/2)/($B$7+A11)					Cell I12=G12-G11						
41	Cell F11=$B$2-(0.0592/2)*LOG(C11/(B11*(10^-$B$8)^3))					Cell J12=A12-A11						
42	Cell G11=F11-$B$9					Cell K12=I12/J12						
43	Cell F24=((2*$B$2+$B$3)/3)-(0.0592/3)*LOG10(1/(10^-$B$8)^3)					Cell L13=K13-K12 (hdden)						
44	Cell G24=F24-$B$9					Cell M13=K13-K12 (hidden)						
45	Cell D25=($B$4*$B$6*2)/($B$7+A25)					Cell N13=H13-H12 (hidden)						
46	Cell E25=($B$5*A25-$B$4*$B$6*2)/($B$7+A25)					Cell O13=M13/N13						
47	Cell F25=$B$3-0.0592*LOG(D25/E25)											
48	Cell G25=F25-$B$9											

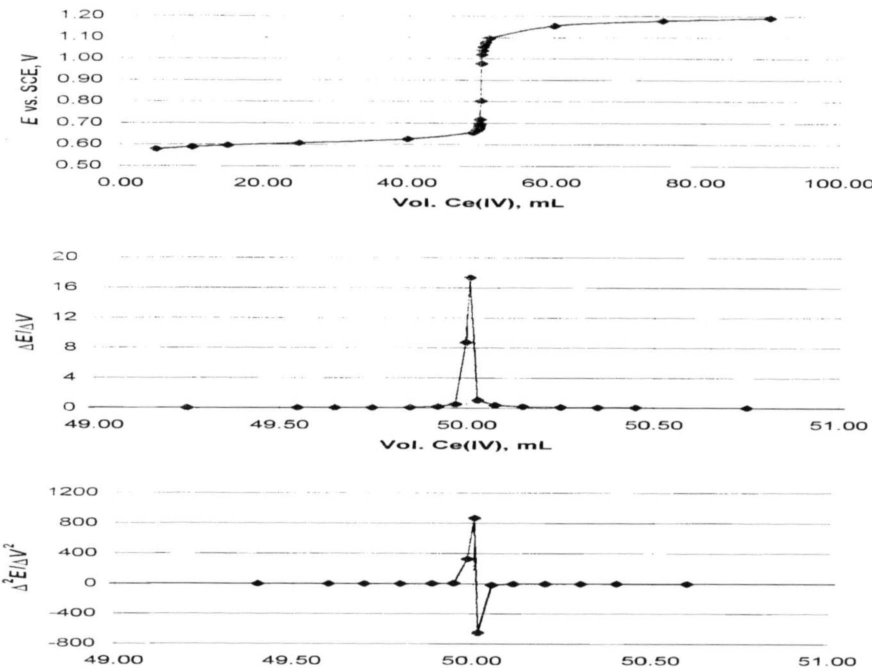

E vs. SCE, V
1.20
1.10
1.00
0.90
0.80
0.70
0.60
0.50
0.00
20.00
40.00
60.00
80.00
100.00
Vol. Ce(IV), mL
ΔE/ΔV
20
16
12
8
4
0
49.00
49.50
50.00
50.50
51.00
Vol. Ce(IV), mL
Δ²E/ΔV²
1200
800
400
0
-400
-800
49.00
49.50
50.00
50.50
51.00
Vol. Ce(IV), mL

19-26. $$\text{pNa} = -\log[\text{Na}^+] = -\left(\frac{E'_{\text{cell}} - K}{0.0592}\right) \qquad \text{where } E'_{\text{cell}} = -0.2331 \text{ V}$$

After addition $E''_{\text{cell}} = -0.1846$ V

$$-\log\left(\frac{10.00 \times [\text{Na}^+] + 1.00 \times 2.00 \times 10^{-2}}{10.00 + 1.00}\right) = -\left(\frac{E''_{\text{cell}} - K}{0.0592}\right)$$

$$-\log(0.9091[\text{Na}^+] + 1.818 \times 10^{-3}) = -\left(\frac{E''_{\text{cell}} - K}{0.0592}\right)$$

Subtracting this latter equation from that for the initial potential gives,

$$-\log[\text{Na}^+] + \log(0.9091[\text{Na}^+] + 1.818 \times 10^{-3}) = -\left(\frac{E'_{\text{cell}} - K}{0.0592}\right) + \left(\frac{E''_{\text{cell}} - K}{0.0592}\right)$$

$$= \frac{E''_{\text{cell}} - E'_{\text{cell}}}{0.0592}$$

$$-\log\left(\frac{[\text{Na}^+]}{0.9091[\text{Na}^+] + 1.818 \times 10^{-3}}\right) = \frac{-0.1846 + 0.2331}{0.0592} = 0.8193$$

or, $$\log\left(\frac{[\text{Na}^+]}{0.9091[\text{Na}^+] + 1.818 \times 10^{-3}}\right) = -0.8193$$

$$\frac{[\text{Na}^+]}{0.9091[\text{Na}^+] + 1.818 \times 10^{-3}} = \text{antilog}(-0.8193) = 0.1516$$

$$[\text{Na}^+] = 0.2478[\text{Na}^+] + 2.756 \times 10^{-4}$$

$$[\text{Na}^+] = 3.197 \times 10^{-4} \text{ M} \quad \text{or} \quad 3.2 \times 10^{-4} \text{ M}$$

19-28. Here we follow the procedure given in the Spreadsheet Exercise on page 163 to determine the slope and intercept. Theoretically for a cation with $n = 1$, the slope given by Equation 19-18 on page 494 should be +0.0592 V, or 59.2 mV. Our slope is slightly higher than this as shown in the spreadsheet that follows, but the plot is linear. Hence, we conclude that the Nernst equation is obeyed with a slightly larger than theoretical slope. We calculate the logarithms of the unknown concentrations from

$$\log c_{unk} = \frac{E_{unk} - \text{intercept}}{\text{slope}}$$

and the concentrations from

$$c_{unk} = 10^{\log c_{unk}}$$

	A	B	C	D	E	F	G	H
1	**19-28 Lithium ion selective electrode**							
2								
3	a_{Li+}	log a_{Li+}	*E* vs. SCE, mV					
4	0.100	-1.00	1.0					
5	0.050	-1.30	-30.0					
6	0.010	-2.00	-60.0					
7	0.001	-3.00	-138.0					
8	Unk 1		-48.5					
9	Unk 2		-75.3					
10								
11	Slope	66.6397						
12	Intercept	64.8846						
13	log Cunk1	-1.7015						
14	Cunk1	0.020						
15	log Cunk2	-2.1036						
16	Cunk 2	0.008						
17	**Documentation**							
18	Cell B11=SLOPE(C4:C7,B4:B7)							
19	Cell B12=INTERCEPT(C4:C7,B4:B7)							
20	Cell B13=(C8-$B$12)/$B$11							
21	Cell B14=10^B13							
22	Cell B15=(C9-$B$12)/$B$11							
23	Cell B16=10^B15							
24								
25								
26								
27								
28								
29								
30								
31								
32								
33								
34								
35								
36								
37								
38								
39								
40								

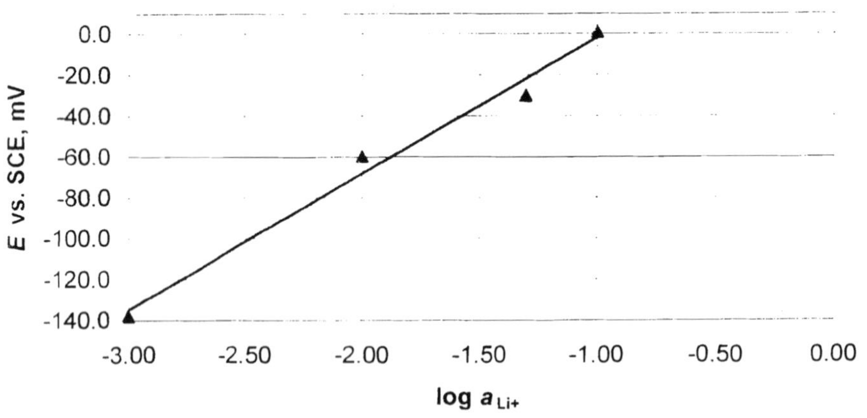

Chapter 20

20-1. **(a)** *Concentration polarization* is a condition in which the current in an electrochemical cell is limited by the rate at which reactants are brought to or removed from the surface of one or both electrodes. *Kinetic polarization* is a condition in which the current in an electrochemical cell is limited by the rate at which electrons are transferred between the electrode surfaces and reactants in solution. For either type of polarization, the current is no longer proportional to the cell potential.

(c) Both the *coulomb* and the *Faraday* are units of quantity of charge, or electricity. The former is the quantity transported by one ampere of current in one second; the latter is equal to 96,495 coulombs or one mole of electrons.

(e) The *electrolysis circuit* consists of a working electrode and a counter electrode. The *control circuit* regulates the applied potential such that the potential between the working electrode and a reference electrode in the control circuit is constant and at a desired level.

20-2. **(a)** *Current density* is the current at an electrode divided by the surface area of that electrode. Ordinarily, it has units of amperes per square centimeter.

(c) A *coulometric titration* is an electroanalytical method in which a constant current of known magnitude generates a reagent that reacts with the analyte. The time required to generate enough reagent to complete the reaction is measured.

(e) *Current efficiency* is a measure of agreement between the number of faradays of current and the number of moles or reactant oxidized or reduced at a working electrode.

20-3. Mass transport in an electrochemical cell results from one or more of the following: (1) *diffusion*, which arises from concentration differences between the electrode surface and the bulk of the solution; (2) *migration*, which results from electrostatic attraction or repulsion; and (3) *convection*, which results from stirring, vibration, or temperature difference.

20-5. Both kinetic and concentration polarization cause the potential of a cell to be more negative than the thermodynamic potential. Concentration polarization arises from the slow rate at which reactants or products are transported to or away from the electrode surfaces. Kinetic polarization arises from the slow rate of the electrochemical reactions at the electrode surfaces.

20-7. Kinetic polarization is often encountered when the product of a reaction is a gas, particularly when the electrode is a soft metal such as mercury, zinc, or copper. It is likely to occur at low temperatures and high current densities.

20-9. Temperature, current density, complexation of the analyte, and codeposition of a gas influence the physical properties of an electrogravimetric deposit.

20-11. (a) An *amperostat* is an instrument that provides a constant current to an electrolysis cell.

(b) A *potentiostat* controls the applied potential to maintain a constant potential between the working electrode and a reference electrode.

20-12. In *amperostatic coulometry*, the cell is operated so that the current in the cell is held constant. In *potentiostatic coulometry*, the potential of the working electrode is maintained constant.

20-13. The species produced at the counter electrode is a potential interference by reacting with the product at the working electrode. Isolation of the one from the other is ordinarily necessary.

20-15. (a) *Voltammetry* is an analytical technique that is based upon measuring the current that develops in a microelectrode as the applied potential is varied. *Polarography* is a particular type of voltammetry in which the microelectrode is a dropping mercury electrode.

20-16. (a) *Voltammograms* are plots of current as a function of voltage applied to a microelectrode.

(e) The *half-wave potential* is the potential on a voltammetric wave at which the current is one half of the limiting current.

20-18. (b)

$$0.020\text{ C} \times \frac{1\text{ F}}{96{,}485\text{ C}} \times \frac{1\text{ mol e}^-}{\text{F}} \times \frac{1\text{ mol cation}}{2\text{ mol e}^-} \times 6.02 \times 10^{23}\,\frac{\text{cation}}{\text{mol cation}} =$$

$$\underline{\underline{6.2 \times 10^{16}\text{ cations}}}$$

20-19. (a) The electrode potential required to begin deposition of copper at a cathode is

$$E_{\text{right}} = 0.337 - \frac{0.0592}{2}\log\frac{1}{0.150} = 0.313\text{ V}$$

The formation of oxygen at the anode requires an electrode potential of

$$E_{\text{left}} = 1.229 - \frac{0.0592}{4}\log\frac{1}{1.00\times(1.00\times10^{-3})^4} = 1.051\text{ V}$$

$$E_{\text{applied}} = E_{\text{right}} - E_{\text{left}} = 0.313 - 1.051 = \underline{\underline{-0.738\text{ V}}}$$

(c) Here, H_2 is formed at the cathode and

$$[H^+] = \text{antilog}(-3.40) = 3.98\times10^{-4}$$

$$E_{\text{right}} = 0.000 - \frac{0.0592}{2}\log\frac{765/760}{(3.98\times10^{-4})^2} = -0.201\text{ V}$$

AgBr is deposited at the anode by the reaction

$$Ag(s) + Br^- \rightarrow AgBr(s) + e^- \quad \text{and}$$

$$E_{\text{left}} = 0.073 - 0.0592\log 0.0864 = 0.136\text{ V}$$

$$E_{\text{applied}} = -0.201 - 0.136 = \underline{\underline{-0.337\text{ V}}}$$

20-20. $E_{\text{right}} = -0.763 - 0.0296\log(1/3.75\times10^{-3}) = -0.835$

$$E_{\text{left}} = -0.277 - 0.0296\log(1/6.40\times10^{-2}) = -0.312$$

$$E_{\text{cell}} = -0.835 - (-0.312) - 0.078\times5 = \underline{\underline{-0.913\text{ V}}}$$

20-21. $E_{\text{right}} = -0.403 - 0.0296\log(1/7.50\times10^{-2}) = -0.436$

$$E_{\text{left}} = -0.136 - 0.0296\log(1/8.22\times10^{-4}) = -0.227$$

$$E_{\text{cell}} = -0.436 - (-0.227) - 0.072\times3.95 = \underline{\underline{-0.493\text{ V}}}$$

20-24. **(a)** $E_{\text{right}} = -0.31 - 0.0592 \log \dfrac{(0.320)^2}{0.150} = -0.30\ \text{V}$

$$E_{\text{left}} = 1.229 - \frac{0.0592}{4} \log \frac{1}{1.00\,(1.00 \times 10^{-10})^4} = 0.637\ \text{V}$$

$$E_{\text{cell}} = -0.30 - 0.637 = \underline{\underline{-0.94\ \text{V}}}$$

(b) $IR = -0.12 \times 2.90 = \underline{\underline{-0.35\ \text{V}}}$

(c) $E_{\text{applied}} = -0.94 - 0.35 - 0.80 = \underline{\underline{-2.09\ \text{V}}}$ [see Solution 20 – 19(c)]

(d) Reduction of 0.150 mol/L $Ag(CN)_2^-$ to Ag^+ produces

$$2 \times 0.150 = 0.300 \text{ mol } CN^-/L \text{ and } [CN^-] \approx 0.300 + 0.320 = 0.620.$$

$$E_{\text{right}} = -0.31 - 0.0592 \log \frac{(0.620)^2}{1.00 \times 10^{-5}} = -0.581\ \text{V}$$

$$E_{\text{left}} = -0.581 - 0.637 - 0.35 - 0.80 = \underline{\underline{-2.37\ \text{V}}}$$

20-25. Cd begins to form when

$$E = -0.403 - 0.0296 \log (1/0.0750) = -0.436\ \text{V}$$

(a) $[Co^{2+}]$ concentration when Cd first forms is given by

$$-0.436 = -0.277 - 0.0296 \log (1/[Co^{2+}])$$

$$\log [Co^{2+}] = (-0.436 + 0.277)/0.0296 = -5.372$$

$$[Co^{2+}] = \text{antilog}(-5.372) = \underline{\underline{4.2 \times 10^{-6}}}$$

(b) $E_{\text{cathode}} = -0.277 - 0.0296 \log (1/1.00 \times 10^{-5}) = \underline{\underline{-0.425\ \text{V}}}$

20-28. **(a)** Br^- separation begins when

$$E_{\text{anode}} = 0.073 - 0.0592 \log 0.250 = 0.109\ \text{V}$$

When $[I^-] = 10^{-5}$

$$E_{anode} = -0.151 - 0.0592 \log 10^{-5} = 0.145 \text{ V}$$

The equilibrium $[Br^-]$ corresponding to this potential is

$$0.145 = 0.073 - 0.0592 \log [Br^-]$$

$$[Br^-] = \text{antilog}\,[(0.073 - 0.145)/0.0592] = 0.061$$

Thus a separation is impossible.

(b) Proceeding in the same way for Cl^-

$$0.145 = +0.222 - 0.0592 \log [Cl^-]$$

$$[Cl^-] = \text{antilog}\,[(0.222 - 0.145)/0.0592] = 20$$

Thus if $[Cl^-] < 20$ M, formation of AgCl will not occur and separation is feasible.

(c) To separate I^- quantitatively

$$E \text{ (vs. SCE)} = 0.244 - 0.145 = 0.099 \text{ V}$$

To prevent Cl^- from precipitating

$$E \text{ (vs. SCE)} = 0.244 - (0.222 - 0.0592 \log 0.250) = -0.014 \text{ V}$$

Thus, a cell consisting of a saturated calomel electrode and the analyte half-cell would be galvanic, and a separation would be feasible by allowing the cell to discharge to 0.00 V.

20-29. Deposition of A is complete when

$$E_A = E_A^0 - \frac{0.0592}{n} \log \frac{1}{1.00 \times 10^{-5}} = E_A^0 - \frac{0.296}{n_A} \text{ V}$$

Deposition of B begins when

$$E_B = E_B^0 - \frac{0.0592}{n}\log\frac{1}{1.00\times10^{-1}} = E_B^0 - \frac{0.0592}{n_B}\text{ V}$$

Boundary condition is that $E_A = E_B$. Thus,

$$E_A^0 - 0.296/n_A = E_B^0 - 0.0592/n_B$$

or $$E_A^0 - E_B^0 = 0.296/n_A - 0.0592/n_B$$

(a) $E_A^0 - E_B^0 = 0.296 - 0.0592 = \underline{\underline{0.237\text{ V}}}$

(c) $E_A^0 - E_B^0 = 0.296/3 - 0.0592 = \underline{\underline{0.0395\text{ V}}}$

(e) $E_A^0 - E_B^0 = 0.296/2 - 0.0592/2 = \underline{\underline{0.118\text{ V}}}$

(g) $E_A^0 - E_B^0 = 0.296/1 - 0.0592/3 = \underline{\underline{0.276\text{ V}}}$

(i) $E_A^0 - E_B^0 = 0.296/3 - 0.0592/3 = \underline{\underline{0.0789\text{ V}}}$

20-30. (a)

$$0.500\text{ g Co}\times\frac{1\text{ mol Co}}{58.93\text{ g Co}}\times\frac{2\text{ mol e}^-}{\text{mol Co}}\times\frac{1\text{ F}}{\text{mol e}^-}\times96485\frac{\text{C}}{\text{F}} = 1.637\times10^3\text{ C}$$

$$1.637\times10^3\text{ C}\times\frac{1\text{ A}\cdot\text{s}}{\text{C}}\times\frac{1}{0.961\text{ A}}\times\frac{1\text{ min}}{60\text{ s}} = \underline{\underline{28.4\text{ min}}}$$

(b) $3Co^{2+} + 4H_2O \rightleftarrows Co_3O_4 + 8H^+ + 2e^-$ $\quad (3/2)\text{ mol }Co^{2+} \equiv 1\text{ mol e}^-$

$$0.500\text{ g Co}\times\frac{1\text{ mol Co}}{58.93\text{ g Co}}\times\frac{1\text{ mol e}^-}{(3/2)\text{ mol Co}}\times\frac{1\text{ F}}{\text{mol e}^-}\times96485\frac{\text{C}}{\text{F}} = 545.8\text{ C}$$

$$545.8\text{ C}\times\frac{1\text{ A}\cdot\text{s}}{\text{C}}\times\frac{1}{0.961\text{ A}}\times\frac{1\text{ min}}{60\text{ s}} = \underline{\underline{9.47\text{ min}}}$$

20-32.

$$(5\times60+24)\text{ s}\times0.401\text{ A}\times\frac{1\text{ C}}{1\text{ A}\cdot\text{s}}\times\frac{1\text{ F}}{96485\text{ C}}\times\frac{1\text{ eq HA}}{\text{F}} = 1.3466\times10^{-3}\text{ eq HA}$$

$$0.1516 \text{ g HA} / 1.3466 \times 10^{-3} \text{ eq HA} \;=\; \underline{\underline{112.6 \text{ g/eq}}}$$

20-34. $1 \text{ mol CaCO}_3 \;\equiv\; 1 \text{ mol HgNH}_3\text{Y}^{2-} \;\equiv\; 2 \text{ mol e}^-$

$$31.6 \times 10^{-3} \text{ A} \times 2.02 \text{ min} \times 60 \frac{\text{s}}{\text{min}} \times \frac{1 \text{ C}}{\text{A} \cdot \text{s}} \times \frac{1 \text{ F}}{96485 \text{ C}} \times \frac{1 \text{ mol e}^-}{1 \text{ F}} \times \frac{1 \text{ mol CaCO}_3}{2 \text{ mol e}^-} \;=$$

$$1.9847 \times 10^{-5} \text{ mol CaCO}_3$$

$$\frac{1.9847 \times 10^{-5} \text{ mol CaCO}_3 \times 100.09 \text{ g CaCO}_3/\text{mol}}{25.00 \text{ mL H}_2\text{O} \times 1.000 \text{ g H}_2\text{O}/\text{mL H}_2\text{O}} \times 10^6 \text{ ppm} \;=\; \underline{\underline{79.5 \text{ ppm CaCO}_3}}$$

20-36. $1 \text{ mol C}_6\text{H}_5\text{NO}_2 \;\equiv\; 4 \text{ mol e}^-$

$$26.74 \text{ C} \times \frac{1 \text{ F}}{96485 \text{ C}} \times \frac{1 \text{ mol e}^-}{1 \text{ F}} \times \frac{1 \text{ mol C}_6\text{H}_5\text{NO}_2}{4 \text{ mol e}^-} \;=\; 6.929 \times 10^{-5} \text{ mol C}_6\text{H}_5 \text{ NO}_2$$

$$\frac{6.929 \times 10^{-5} \text{ mol Ca}_6\text{H}_5\text{NO}_2 \times 123.11 \text{ g C}_6\text{H}_5\text{NO}_2/\text{mol}}{210 \text{ mg sample} \times 10^{-3} \text{ g/mg}} \times 100\,\% \;=\; \underline{\underline{4.06\% \text{ C}_6\text{H}_5\text{NO}_2}}$$

20-38. $1 \text{ mol CCl}_4 \;\equiv\; 1 \text{ mol e}^- \qquad 1 \text{ mol CHCl}_3 \;\equiv\; 3 \text{ mol e}^-$

$$11.63 \text{ C} \times \frac{1 \text{ F}}{96485 \text{ C}} \times \frac{1 \text{ mol e}^-}{\text{F}} \times \frac{1 \text{ mol CCl}_4}{\text{mol e}^-} \;=\; 1.205 \times 10^{-4} \text{ mol CCl}_4$$

$$\frac{68.6}{96485} \text{ F} \times \frac{1 \text{ mol e}^-}{\text{F}} \times \frac{1 \text{ mol CHCl}_3}{3 \text{ mol e}^-} \;=\; 2.370 \times 10^{-4} \text{ mol CHCl}_3$$

$$\text{original no. mol CHCl}_3 \;=\; 2.370 \times 10^{-4} - 1.205 \times 10^{-4} \;=\; 1.165 \times 10^{-4}$$

$$\frac{1.205 \times 10^{-4} \text{ mol CCl}_4 \times 153.82 \text{ g CCl}_4/\text{mol CCl}_4}{0.750 \text{ g sample}} \times 100\,\% \;=\; \underline{\underline{2.471\% \text{ CCl}_4}}$$

$$\frac{1.165 \times 10^{-4} \text{ mol CHCl}_3 \times 119.38 \text{ g CHCl}_3/\text{mol}}{0.750 \text{ g sample}} \times 100\,\% \;=\; \underline{\underline{1.854\% \text{ CHCl}_3}}$$

20-40.

$$\text{mol Cd deposited} = i_1\ (\text{A}) \times t\ (\text{min}) \times \left(\frac{60\ \text{s}}{\text{min}}\right) \times \left(\frac{1\ \text{C}}{\text{A}\cdot\text{s}}\right) \times \frac{1}{F}\left(\frac{\text{mol e}^-}{\text{C}}\right) \times \frac{1}{2}\left(\frac{\text{mol Cd deposited}}{\text{mol e}^-}\right)$$

$$\%\ \text{decrease in}\ [\text{Cd}^{2+}] = \frac{\text{mol Cd deposited}}{\text{mol Cd}^{2+}\ \text{initially present}} \times 100\%$$ (see spreadsheet)

	A	B	C
1	**20-40 Voltammogram for Cd^{2+}**		
2	Limiting current, A	3.13E-05	
3	Faraday, C mol^{-1}	96485	
4	Conc. Cd^{2+}, M	3.65E-03	
5	Volume solution, mL	20.00	
6	Initial mol Cd^{2+} present	7.30E-05	
7	**Electrolysis time, min**	**mol Cd deposited**	**% decrease in [Cd^{2+}]**
8	5	4.87E-08	0.067
9	10	9.73E-08	0.13
10	30	2.92E-07	0.40
11			
12	**Documentation**		
13	Cell B8=$B$2*A8*60/(2*$B$3)		
14	Cell C8=(B8/$B$6)*100		

20-42. Stripping methods are generally more sensitive than other voltammetric procedures because the analyte can be removed from a relatively large volume of solution and concentrated on a tiny electrode for a long period (often many minutes). After concentration, the potential is reversed and all of the analyte that has been stored on the electrode is rapidly oxidized or reduced producing a large current.

Chapter 21

21-1. A solution of $Cu(NH_3)_4^{2+}$ is blue because this ion absorbs yellow radiation and transmits blue radiation unchanged.

21-2. **(a)** Absorbance A and transmittance T are related by the equation

$$A = -\log T = \log\frac{1}{T}$$

21-3. Failure to use monochromatic radiation, existence of stray radiation, experimental uncertainties in measurement of low absorbances, molecular interactions at high absorbance, concentration-dependent association/dissociation.

21-6.
$$\nu = c/\lambda = 3.00\times10^{10}\text{ cm s}^{-1}/\lambda\text{ (cm)} = (3.00\times10^{10}/\lambda)\text{ s}^{-1}$$
$$= (3.00\times10^{10}/\lambda)\text{ Hz}$$

(a) $$\nu = 3.00\times10^{10}\text{ cm s}^{-1}/(2.65\text{ Å}\times10^{-8}\text{ cm/Å}) = \underline{\underline{1.13\times10^{18}\text{ Hz}}}$$

(c) $$\nu = 3.00\times10^{10}\text{ cm s}^{-1}/(694.3\text{ nm}\times10^{-7}\text{ cm/nm}) = \underline{\underline{4.318\times10^{14}\text{ Hz}}}$$

(e) $$\nu = 3.00\times10^{10}\text{ cm s}^{-1}/(19.6\ \mu\text{m}\times10^{-4}\text{ cm/}\mu\text{m}) = \underline{\underline{1.53\times10^{13}\text{ Hz}}}$$

21-7. **(a)** $$\lambda = \frac{c}{\nu} = \frac{3.00\times10^{10}\text{ cm s}^{-1}}{118.6\text{ MHz}\times10^{6}\text{ Hz/MHz}}\times\frac{1\text{ Hz}}{\text{s}^{-1}} = \underline{\underline{252.8\text{ cm}}}$$

(c) $$\lambda = \frac{c}{\nu} = \frac{3.00\times10^{10}\text{ cm s}^{-1}}{105\text{ MHz}\times10^{6}\text{ Hz/MHz}}\times\frac{1\text{ Hz}}{\text{s}^{-1}} = \underline{\underline{286\text{ cm}}}$$

21-9. $$\bar{\nu} = 1/(185\text{ nm}\times10^{-7}\text{ cm/nm}) = \underline{\underline{5.41\times10^{4}\text{ cm}^{-1}}}\text{ to}$$
$$1/3000\times10^{-7} = \underline{\underline{3.33\times10^{3}\text{ cm}^{-1}}}$$

$$\nu = 3.00\times10^{10}\text{ cm s}^{-1}\times5.41\times10^{4}\text{ cm}^{-1} = \underline{\underline{1.62\times10^{15}\text{ Hz}}}\text{ to}$$
$$3.00\times10^{10}\times3.33\times10^{3} = \underline{\underline{1.00\times10^{14}\text{ Hz}}}$$

21-11.

$$\lambda = \frac{c}{\nu} = \frac{3.00\times10^{10}\text{ cm s}^{-1}}{220\times10^{6}\text{ s}^{-1}} = \underline{\underline{136\text{ cm}}}$$

$$E = h\nu = 6.63\times10^{-34}\text{ J s}\times220\times10^{6}\text{ s}^{-1} = \underline{\underline{1.46\times10^{-25}\text{ J}}}$$

21-12. (a) The frequency ν of the radiation is

$$\nu = \frac{3.00\times10^{10}\text{ cm/s}}{589\text{ nm}\times10^{-7}\text{ cm/nm}} = 5.093\times10^{14}\text{ s}^{-1} \qquad \text{(Equation 21} - 2)$$

The velocity of radiation in the aqueous solution v can be computed from the refractive index (see the Margin Note in text page 549).

$$\text{refractive index} = \eta = 3.00\times10^{10}/v = 1.35$$

$$v = 3.00\times10^{10}\text{ cm s}^{-1}/1.35 = 2.222\times10^{10}\text{ cm s}^{-1}$$

Substituting into Equation 21-1 after rearranging gives

$$\lambda = v/\nu = \frac{2.222\times10^{10}\text{ cm s}^{-1}}{5.093\times10^{14}\text{ s}^{-1}}\times\frac{10^{7}\text{ nm}}{\text{cm}} = \underline{\underline{436\text{ nm}}}$$

21-13. (a) cm^{-1} ppm^{-1} **(c)** cm^{-1} %$^{-1}$

21-14. (a) $T = 100\times\text{antilog}(-0.0510) = \underline{\underline{88.9\%}}$

Proceeding in the same way, we obtain

(c) $\underline{\underline{41.8\%}}$ **(e)** $\underline{\underline{32.7\%}}$

21-15. (a) $A = -\log(25.5/100) = \underline{\underline{0.593}}$

Proceeding in the same way, we obtain

(c) $\underline{\underline{0.484}}$ **(e)** $\underline{\underline{1.07}}$

21-18. (a) $T = (\text{antilog} - 0.172)\times100\,\% = \underline{\underline{67.3\%}}$

$$c = \frac{0.172}{4.23\times10^{3}\text{ L cm}^{-1}\text{ mol}^{-1}\times1.00\text{ cm}} = 4.066\times10^{-5} = \underline{\underline{4.07\times10^{-5}\text{ M}}}$$

$$c_{ppm} = 4.066\times10^{-5}\frac{\text{mol}}{\text{L}}\times\frac{200\text{ g}}{\text{mol}}\times\frac{10^3\text{ mg}}{\text{g}}\times\frac{1\text{ ppm}}{1\text{ mg/L}} = \underline{\underline{8.13\text{ ppm}}}$$

$$a = A/(b\times c_{ppm}) = 0.172/(1.00\times8.14) = \underline{\underline{2.11\times10^{-2}\text{ cm}^{-1}\text{ ppm}^{-1}}}$$

(c) Proceeding as in part (a), we obtain

$$\underline{\underline{T = 30.2\%}}, \qquad \underline{\underline{c = 6.54\times10^{-5}\text{ M}}},$$

$$\underline{\underline{c_{ppm} = 13.1\text{ ppm}}}, \qquad \underline{\underline{a = 0.0397\text{ cm}^{-1}\text{ ppm}^{-1}}}$$

(e) $A = \varepsilon bc = 3.73\times10^3\times1.71\times10^{-3}\times0.100 = \underline{\underline{0.638}}$

$$T = (\text{antilog}-0.638)\times100\% = \underline{\underline{23.0\%}}$$

$$c_{ppm} = 1.71\times10^{-3}\frac{\text{mol}}{\text{L}}\times200\times10^3\frac{\text{mg}}{\text{mol}} = 342\frac{\text{mg}}{\text{L}} = \underline{\underline{342\text{ ppm}}}$$

$$a = 0.638/(0.100\text{ cm}\times342\text{ ppm}) = \underline{\underline{1.87\times10^{-2}\text{ cm}^{-1}\text{ ppm}^{-1}}}$$

(g) $T = \text{antilog}\,(0.798)\times100\% = \underline{\underline{15.9\%}}$

$$c = 33.6\frac{\text{mg}}{\text{L}}\times\frac{1\text{ mol}}{200\times10^3\text{ mg}} = \underline{\underline{1.68\times10^{-4}\frac{\text{mol}}{\text{L}}}}$$

$$\varepsilon = 0.798/(1.50\text{ cm}\times1.68\times10^{-4}\text{ mol/L}) = \underline{\underline{3.17\times10^3\text{ L cm}^{-1}\text{ mol}^{-1}}}$$

$$a = 0.798/(1.50\text{ cm}\times33.6\text{ ppm}) = \underline{\underline{0.0158\text{ cm}^{-1}\text{ mol}^{-1}}}$$

(i)

$$c = 5.24\frac{\text{mg}}{\text{L}}\times\frac{1\text{ mol}}{200\times10^3\text{mg}} = \underline{\underline{2.62\times10^{-5}\frac{\text{mol}}{\text{L}}}}$$

$$A = -\log0.0523 = \underline{\underline{1.281}}$$

$$b = \frac{A}{\varepsilon c} = \frac{1.281}{9.78\times10^3\times2.62\times10^{-5}} = \underline{\underline{5.00\text{ cm}}}$$

$$a = \frac{1.281}{5.00\text{ cm} \times 5.24\text{ ppm}} = \underline{\underline{0.0489\text{ cm}^{-1}\text{ ppm}^{-1}}}$$

21-19.

$$c = \frac{4.48\text{ mg KMnO}_4}{\text{L}} \times \frac{10^{-3}\text{ g KMnO}_4}{\text{mg KMnO}_4} \times \frac{1\text{ mol KMnO}_4}{158.03\text{ g KMnO}_4}$$

$$= 2.835 \times 10^{-5}\,\frac{\text{mol KMnO}_4}{\text{L}}$$

$$A = -\log 0.309 = 0.5100$$

$$\varepsilon = \frac{0.5100}{1.00 \times 2.835 \times 10^{-5}} = \underline{\underline{1.80 \times 10^4\text{ L cm}^{-1}\text{ mol}^{-1}}}$$

21-21. **(a)** $A_1 = \varepsilon b c = 7.00 \times 10^3 \times 1.00 \times 2.50 \times 10^{-5} = \underline{\underline{0.175}}$

(b) $A_2 = 2 \times A_1 = 2 \times 0.175 = \underline{\underline{0.350}}$

(c) $T_1 = \text{antilog}\,(-0.175) = \underline{\underline{0.668 \quad \text{or} \quad 66.8\%}}$

$T_2 = \text{antilog}\,(-0.350) = \underline{\underline{0.447 \quad \text{or} \quad 44.7\%}}$

(d) $A = -\log\,(0.668/2) = \underline{\underline{0.476}}$

21-22. $\varepsilon = 7.00 \times 10^3\text{ L cm}^{-1}\text{ mol}^{-1}$

$$c_{\text{Fe}} = 3.38\,\frac{\text{mg Fe}}{\text{L}} \times 10^{-3}\,\frac{\text{g Fe}}{\text{mg Fe}} \times \frac{1\text{ mol Fe}}{55.85\text{ g Fe}} \times \frac{2.50\text{ mL}}{50.0\text{ mL}} = 3.026 \times 10^{-6}\text{ M}$$

$$A = 7.00 \times 10^3 \times 2.50 \times 3.026 \times 10^{-6} = \underline{\underline{0.0530}}$$

21-24. $\varepsilon = 7.00 \times 10^3\text{ L cm}^{-1}\text{ mol}^{-1}$

(a) $A = \varepsilon b C = 7.00 \times 10^3 \times 1.00 \times 8.50 \times 10^{-5} = \underline{\underline{0.595}}$

(b) $T = (\text{antilog} - 0.595) \times 100\% = \underline{\underline{25.4\%}}$

(c) $c = A/\varepsilon b = 0.595/(7.00 \times 10^3 \times 5.00) = \underline{\underline{1.70 \times 10^{-5}\ \mathrm{M}}}$

(d) $b = A/\varepsilon c = 0.595/(7.00 \times 10^3 \times 3.40 \times 10^{-5}) = \underline{\underline{2.50\ \mathrm{cm}}}$

21-26.

	A	B	C	D
1	**Problem 21-26 Determination of compound X**			
2	Concentration in ppm	Measured absorbance		
3	0.5	0.24		
4	1.5	0.36		
5	2.5	0.44		
6	3.5	0.59		
7	4.5	0.70		
8	unknown	0.50		
9	**Regression equation**			
10	Slope	0.12		
11	Intercept	0.18		
12	Concentration of unknown	2.8		
13	**Error analysis**			
14	s_r (standard error in y)	0.017029		
15	N	5		
16	S_{xx}	10		
17	y bar (average absorbance)	0.466		
18	M	1		
19	s_c (standard deviation in c)	0.2		
20	**Spreadsheet Documentation**			
21	Cell B10=SLOPE(B3:B7,A3:A7)			
22	Cell B11=INTERCEPT(B3:B7,A3:A7)			
23	Cell B12=(B8-B11)/B10			
24	Cell B14=STEYX(B3:B7,A3:A7)			
25	Cell B15=COUNT(B3:B7)			
26	Cell B16=B15*VARP(A3:A7)			
27	Cell B17=AVERAGE(B3:B7)			
28	Cell B19=B14/B10*SQRT(1/B18+1/B15+((B8-B17)^2)/((B10^2)*B16))			
29				
30				
31				
32				
33				
34				
35				
36				
37				
38				
39				
40				
41				

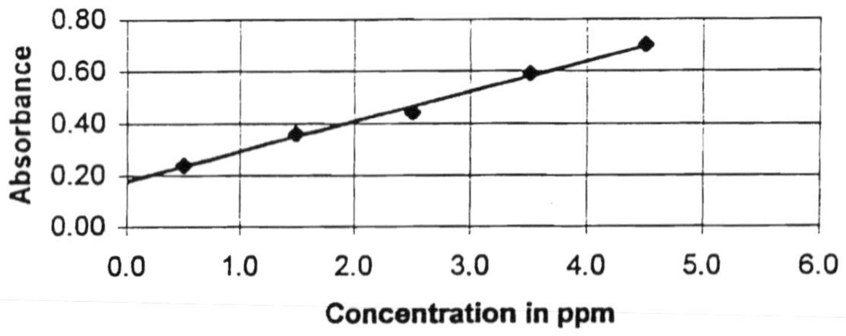

Chapter 22

22-1. **(b)** A *phototube* is a vacuum tube equipped with a photoemissive cathode. When photons strike the photocathode, photoelectrons are emitted by the photoelectric effect. When a voltage of 90 V or more is applied between cathode and anode, the photoelectrons are attracted to the anode and collected to give a small photocurrent. A *photoconductive cell* consists of a thin film of a semiconductor material. Absorption of radiation decreases the electrical resistance of the semiconductor. When placed in series with a voltage source and load resistor, the voltage drop across the load resistor is measured.

(d) A diode array spectrometer detects the entire spectral range simultaneously and can produce a spectrum in less than one second. Conventional spectrometers must scan the spectrum by mechanically rotating a grating. Consequently they take minutes to obtain the spectrum. Diode array instruments can be used to obtain spectra when changes are occurring rapidly such as the output of a liquid chromatograph or a fast kinetics apparatus.

Conventional instruments are normally too slow for such tasks.

22-3. Photons from the infrared region of the spectrum do not have sufficient energy to cause photoemission from the cathode of a photomultipler.

22-5. *Tungsten/halogen lamps* contain a small amount of iodine in the evacuated quartz envelope that contains the tungsten filament. The iodine prolongs the life of the lamp and permits it to operate at a higher temperature. The iodine combines with gaseous tungsten that sublimes from the filament and causes the metal to be redeposited, thus adding to the life of the lamp.

22-7. As a minimum, the radiation emitted by the source of a single-beam instrument must be stable for however long it takes to make the 0% *T* adjustment, the 100% *T* adjustment, and the measurement of *T* for the sample.

22-9. pH, electrolyte concentration, temperature.

22-11. **(a)** $T = 179\ \text{mV}/685\ \text{mV} = 0.261$ or $\underline{\underline{26.1\%}}$

$$A = -\log 0.261 = \underline{\underline{0.583}}$$

(c) $A = 2 \times 0.583 = \underline{\underline{1.166}}$

$$T = \text{antilog}(-1.166) = \underline{\underline{0.0682}} \text{ or } \underline{\underline{6.82\%}}$$

22-12. (b) $A = -\log 0.338 = \underline{\underline{0.471}}$

(d) $A = 2 \times 0.471 = 0.942$ and $T = \text{antilog}(-0.942) = \underline{\underline{0.114}}$

	A	B	C	D	E	F	G
1	22-13 Spreadsheet of data taken from diode array spectrophotometer						
2							
3		Wavelength, nm	$P_{solvent}$	$P_{solution}$	T	A	
4		350	0.002689	0.002560	0.952357	0.021	
5		375	0.006326	0.005995	0.947545	0.023	
6		400	0.016975	0.015143	0.892072	0.050	
7		425	0.035517	0.031648	0.891046	0.050	
8		450	0.062425	0.024978	0.400129	0.398	
9		475	0.095374	0.019073	0.199986	0.699	
10		500	0.140567	0.023275	0.165577	0.781	
11		525	0.188984	0.037448	0.198153	0.703	
12		550	0.263103	0.088537	0.336512	0.473	
13		575	0.318361	0.200872	0.630957	0.200	
14		600	0.394600	0.278072	0.704693	0.152	
15		625	0.477018	0.363525	0.762079	0.118	
16		650	0.564295	0.468281	0.829851	0.081	
17		675	0.655066	0.611062	0.932825	0.030	
18		700	0.739180	0.704126	0.952577	0.021	
19		725	0.813694	0.777466	0.955476	0.020	
20		750	0.885979	0.863224	0.974316	0.011	
21		775	0.945083	0.921446	0.97499	0.011	
22		800	1.000000	0.977237	0.977237	0.010	
23		Documentation					
24		Cell E4=D4/C4					
25		Cell F4=-LOG10(E4)					
26							

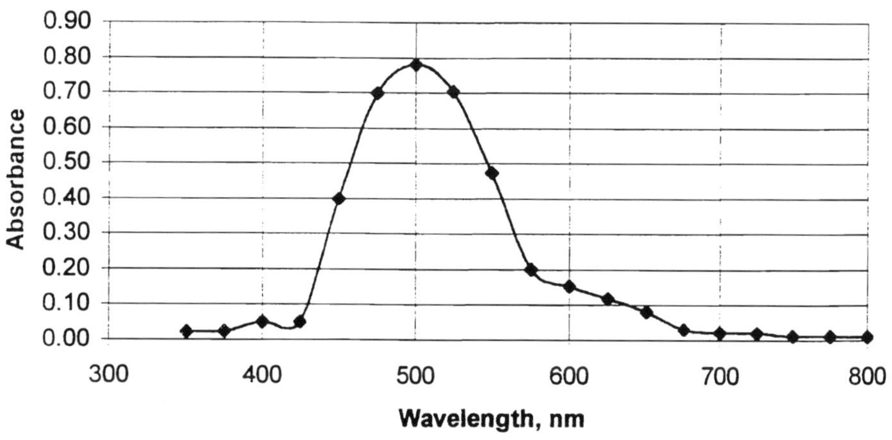

Chapter 23

23-1. **(a)** A *chromophore* is an organic functional group that absorbs radiation in the ultraviolet/visible regions.

(c) Radiation that consists of a single wavelength is said to be *monochromatic*.

(e) An *absorption spectrum* is a plot of a spectral property (absorbance, log absorbance, absorptivity, transmittance) as the ordinate and wavelength, wavenumber, or frequency as the abscissa.

23-2. **(a)** *Fluorescence* is a process by which an excited singlet species relaxes by emitting electromagnetic radiation.

(c) *Internal conversion* is the nonradiative relaxation of a molecule from the lowest vibrational level of an excited electronic state to the highest vibrational level of a lower electronic state.

(e) The *Stokes shift* is the difference in wavelength between the radiation used to excite fluorescence and the wavelength of the emitted radiation.

(g) An *inner-filter effect* occurs when the fluorescent radiation from an excited analyte molecule is absorbed by an unexcited analyte molecule. This process results in a decrease in fluorescence intensity. Also, excessive absorption of the incident beam causes a primary inner-filter effect.

23-4. Compounds that fluoresce have structures that slow the rate of nonradiative relaxation to the point where there is time for fluorescence to occur. Compounds that do not fluoresce have structures that permit rapid relaxation by nonradiative processes.

23-6. **(a)** Excitation of fluorescence usually involves transfer of an electron to a high vibrational state of an upper electronic state. Relaxation to a lower vibrational state of this electronic state goes on much more rapidly than fluorescence relaxation. When fluorescence relaxation occurs it is to a high vibrational state of the ground state or to a high vibrational state of an electronic state that is above the ground state. Such transitions involve less energy than the excitation energy. Therefore, the emitted radiation is longer in wavelength than the excitation wavelength.

(b) For spectrofluorometry, the analytical signal F is given by $F = 2.3\,K'\,\varepsilon bc\,P_0$. The magnitude of F, and thus sensitivity, can be enhanced by increasing the source intensity P_0 or the transducer sensitivity.

For spectrophotometry, the analytical signal A is given by $A = \log P_0/P$. Increasing P_0 is accompanied by a corresponding increase in P. Thus, the ratio does not change nor does the analytical signal. Consequently, no improvement in sensitivity accompanies such changes.

23-8. Charge-transfer absorption occurs in species that contain both an electron donor and an electron acceptor group. The absorbed energy results in an electron being transferred from an orbital of the donor group to an orbital that is largely associated with the acceptor group. This type of absorption is analytically important because the molar absorptivities associated with this type of transition are usually very high, which leads to high sensitivities and low detection limits.

23-10. In *atomic emission spectroscopy* the radiation source is the sample itself. The energy for excitation of analyte atoms comes from a flame, a furnace, or a plasma. The signal is the measured intensity of the flame at the wavelength of interest. In *atomic absorption spectroscopy* the radiation source is usually a line source such as a hollow cathode lamp, and the output signal is the absorbance calculated from the incident power of the source and the resulting power after the light has passed through the atomized sample in the heated source.

23-11. In atomic emission spectroscopy, the analytical signal is produced by *excited* atoms or ions, whereas in atomic absorption the signal results from absorption by *unexcited* species. Typically, the number of unexcited species exceeds the excited by several orders of magnitude. The ratio of unexcited to excited atoms in a hot medium varies exponentially with temperature. Thus a small change in temperature brings about a large change in the number of excited atoms. The number of unexcited atoms changes very little, however, because they are present in an enormous excess. Therefore, emission spectroscopy can be more sensitive to temperature changes than is absorption spectroscopy.

23-13. $[H_3O^+][In^-]/[HIn] = 8.00 \times 10^{-5}$

$$[H_3O^+] = [In^-] \qquad [HIn] = c_{HIn} - [In^-] \approx c_{HIn}$$

(a) At $c_{HIn} = 3.00 \times 10^{-4}$

$$\frac{[In^-]^2}{(3.00 \times 10^{-4}) - [In^-]} = 8.00 \times 10^{-5}$$

$$[In^-]^2 + 8.00 \times 10^{-5}[In^-] - 2.40 \times 10^{-8} = 0$$

$$[In^-] = 1.20 \times 10^{-4}$$

$$[HIn] = 3.00 \times 10^{-4} - 1.20 \times 10^{-4} = 1.80 \times 10^{-4}$$

$$A_{430} = 1.20 \times 10^{-4} \times 0.775 \times 10^3 + 1.80 \times 10^{-4} \times 8.04 \times 10^3 = \underline{\underline{1.54}}$$

$$A_{600} = 1.20 \times 10^{-4} \times 6.96 \times 10^3 + 1.80 \times 10^{-4} \times 1.23 \times 10^3 = \underline{\underline{1.06}}$$

The remaining calculations are performed in the same way and yield the following results.

c_{ind}, M	$[In^-]$	$[HIn]$	A_{430}	A_{600}
3.00×10^{-4}	1.20×10^{-4}	1.80×10^{-4}	1.54	1.06
2.00×10^{-4}	9.27×10^{-5}	1.07×10^{-4}	0.935	0.777
1.00×10^{-4}	5.80×10^{-5}	4.20×10^{-5}	0.383	0.455
0.500×10^{-4}	3.48×10^{-5}	1.52×10^{-5}	0.149	0.261
0.250×10^{-4}	2.00×10^{-5}	5.00×10^{-6}	0.056	0.145

23-15. $A = \varepsilon bc = 9.32 \times 10^3 \times 1.00 \times c$

$$c \geq A/9.32 \times 10^3 = 0.15/9.32 \times 10^5 = \underline{\underline{1.6 \times 10^{-5}\text{ M}}}$$

$$c \leq 0.80/9.32 \times 10^{3} = \underline{\underline{8.6 \times 10^{-5}\ \text{M}}}$$

23-17. $A = -\log T \quad A_{10\%} = -\log 0.100 = 1.00 \qquad A_{90\%} = -\log 0.900 = 0.0458$

$\log \varepsilon = 2.75$ and $\varepsilon = 562$

$$c_{10\%} = \frac{A}{b\,\varepsilon} = \frac{1.00}{(1.5)(562)} = \underline{\underline{1.2 \times 10^{-3}\ \text{M}}}$$

$$c_{90\%} = \frac{0.0458}{(1.5)(562)} = \underline{\underline{5.4 \times 10^{-5}\ \text{M}}}$$

23-21. The absorbance will decrease linearly before the equivalence point and reach a constant value of approximately zero after the equivalence point.

23-23. For the unknown alone, we can write Beer's law in the form

$A_x = \varepsilon b c_x V_x / V_t$ where V_t is the total volume of solution

For the solution after standard addition

$$A_s = \varepsilon b\,(c_x V_x + c_s V_s)/V_t$$

Dividing the first equation by the second gives

$$\frac{A_x}{A_s} = \frac{c_x V_x}{c_x V_x + c_s V_s}$$

$$A_x c_x V_x + A_x c_s V_s = A_s c_x V_x$$

This equation rearranges to

$$c_x (A_s V_x - A_x V_x) = A_x c_s V_s$$

$$c_x = \frac{A_x c_s V_s}{V_x (A_s - A_x)}$$

Substituting numerical data gives

$$c_x = \frac{0.398 \times 3.00 \times 5.00}{25.00\,(0.510 - 0.398)} = 2.132 \text{ ppm Co}$$

$$\frac{22.132 \text{ g Co}}{10^6 \text{ g soln}} \times 500 \text{ g soln} = 1.066 \times 10^{-3} \text{ g Co}$$

$$\text{percent Co} = \frac{1.066 \times 10^{-3} \text{ g Co}}{4.97 \text{ g sample}} \times 100\% = \underline{\underline{0.0214\%}}$$

23-24. **(a)** No absorbance until the equivalence point after which the absorbance increases linearly.

(c) Absorbance increases linearly until the equivalence point after which it stays relatively constant.

(e) Absorbance increases linearly until the equivalence point after which it also increases linearly but with a larger slope.

23-26. $A_{365} = \varepsilon_{\text{Co(365)}} \times b \times c_{\text{Co}} + \varepsilon_{\text{Ni(365)}} \times b \times c_{\text{Ni}}$

$A_{700} = \varepsilon_{\text{Co(700)}} \times b \times c_{\text{Co}} + \varepsilon_{\text{Ni(700)}} \times b \times c_{\text{Ni}}$

Since, $\varepsilon_{\text{Ni(700)}} = 0$, $\quad A_{700} = \varepsilon_{\text{Co(700)}} \times b \times c_{\text{Co}}$

$$c_{\text{Co}} = \frac{A_{700}}{\varepsilon_{\text{Co(700)}} \times b} \quad \text{and} \quad c_{\text{Ni}} = \frac{A_{365} - \varepsilon_{\text{Co(365)}} b c_{\text{Co}}}{\varepsilon_{\text{Ni(365)}}}$$

The results are shown in the spreadsheet that follows.

23-26.

	A	B	C	D	E
1	**23-26 Spreadsheet for simultaneous determination of cobalt and nickel**				
2	*Known parameters*	$\varepsilon_{Co(365)}b$ in L mol^{-1}	$\varepsilon_{Co(700)}b$ in L mol^{-1}	$\varepsilon_{Ni(365)}b$ in L mol^{-1}	$\varepsilon_{Ni(700)}b$ in L mol^{-1}
3		3529	3228	428.9	0
4					
5	**Solution**	A_{700}	A_{365}	c_{Co}	c_{Ni}
6	1	0.0235	0.617	5.48E-05	1.31E-04
7	2	0.0714	0.755	1.66E-04	5.19E-05
8	3	0.0945	0.920	2.20E-04	4.41E-05
9	4	0.0147	0.592	3.43E-05	1.46E-04
10	5	0.0540	0.685	1.26E-04	7.46E-05
11					
12	**Documentation**				
13	Cell D6=B6/$D$3				
14	Cell E6=(C6-$B$3*D6)/$C$3				

	A	B	C	D	E
1	**23-27 Spreadsheet for simultaneous determination of cobalt and nickel**				
2	*Known parameters*	$\varepsilon_{Co(510)}b$ in L mol^{-1}	$\varepsilon_{Co(656)}b$ in L mol^{-1}	$\varepsilon_{Ni(510)}b$ in L mol^{-1}	$\varepsilon_{Ni(656)}b$ in L mol^{-1}
3		36400	1240	5520	17500
4					
5	Wt. sample, g	0.425			
6	Diluted volume, mL	50.00			
7	Aliquot size, mL	25.00			
8	Final volume, mL	50.00			
9	M_{Co}	58.93			
10	M_{Ni}	58.69			
11	**By matrix method**				
12	Coefficient Matrix		Constant Matrix		
13	36400	5520	0.446		
14	1240	17500	0.326		
15					
16	Inverse Matrix		Solution Matrix		
17	2.77709E-05	-8.760E-06	9.53E-06		
18	-1.96777E-06	5.776E-05	1.80E-05		
19					
20	**By simult. eqns.**				
21	c_{Co}	9.53E-06			
22	c_{Ni}	1.80E-05			
23					
24	ppm Co	132			
25	ppm Ni	249			
26					
27	**Documentation**				
28	Cells A13:B14=MINVERSE(A13:B14) entered as an array with control+shift+enter				
29	Cells C13:C14=MMULT(A17:B18,C13:C14) entered as an array with control+shift+enter				
30	Cell B21=(D3*C14-E3*C13)/(D3*C3-E3*B3)				
31	Cell B22=(C3*C13-B3*C14)/(D3*C3-B3*E3)				
32	Cell B24=$B$8*C17*($B$6/$B$7)*($B$9/1000)*1000000/$B$5				
33	Cell B25=$B$8*C18*($B$6/$B$7)*($B$9/1000)*1000000/$B$5				

23-29. In Section 12F (page 298), we find

$$\alpha_0 = \frac{[H_3O^+]}{[H_3O^+] + K_{HIn}} \qquad \text{(Equation 12 – 16)}$$

$$\alpha_1 = 1 - \alpha_0$$

At 450 nm and $b = 1.00$

$$A_{450} = \varepsilon_{HIn} \times 1.00 \times [HIn] + \varepsilon_{In^-} \times 1.00 \times [In^-]$$

$$= \varepsilon_{HIn}\,\alpha_0\,c_{In} + \varepsilon_{In^-}\,\alpha_1\,c_{In}$$

$$= (\varepsilon_{HIn}\,\alpha_0 + \varepsilon_{In}\,\alpha_1)\,c_{In}$$

where c_{In} is the analytical concentration of the indicator ($c_{In} = [HIn] + [In^-]$). We may assume at pH 1.00 all of the indicator is present as HIn; at pH 13.0 it is all present as In^-. Therefore from the data in Problem 23-28, we may write

$$\varepsilon_{HIn} = \frac{A_{450}}{b\,c_{In}} = \frac{0.658}{1.00 \times 8.00 \times 10^{-5}} = 8.225 \times 10^3$$

$$\varepsilon_{In^-} = \frac{A_{450}}{b\,c_{In}} = \frac{0.076}{1.00 \times 8.00 \times 10^{-5}} = 9.5 \times 10^2$$

(a) pH $= 4.92 \quad [H_3O^+] = 1.20 \times 10^{-5}$

$$\alpha_0 = \frac{1.20 \times 10^{-5}}{1.20 \times 10^{-5} + 4.80 \times 10^{-6}} = 0.714$$

$$\alpha_1 = 1.000 - 0.714 = 0.286$$

$$A_{450} = (8.225 \times 10^3 \times 0.714 + 9.5 \times 10^2 \times 0.286)\,8.00 \times 10^{-5} = \underline{\underline{0.492}}$$

Proceeding in the same way, we obtain

	pH	$[H_3O^+]$	α_0	α_1	A_{450}
(a)	4.92	1.20×10^{-5}	0.714	0.286	0.492
(c)	5.93	1.18×10^{-6}	0.197	0.803	0.190

23-30. The approach is identical to that of Solution 23-29. At 595 nm and

at pH = 1.00, $$\varepsilon_{HIn} = \frac{A_{595}}{b\,c_{In}} = \frac{0.032}{(1.00)(8.00 \times 10^{-5}\text{ M})} = \underline{\underline{4.0 \times 10^2}}$$

at pH = 13.00, $$\varepsilon_{In^-} = \frac{0.361}{(1.00)(8.00 \times 10^{-5}\text{ M})} = \underline{\underline{4.51 \times 10^3}}$$

(a) At pH = 5.30 and with 1.00-cells, $[H_3O^+] = 5.01 \times 10^{-6}$ and

$$\alpha_0 = \frac{[H_3O^+]}{[H^+] + K_{HIn}} = \frac{5.01 \times 10^{-6}}{5.01 \times 10^{-6} + 4.80 \times 10^{-6}} = 0.511$$

$$\alpha_1 = 1 - \alpha_0 = 0.489$$

$$A_{595} = (\varepsilon_{HIn}\alpha_0 + \varepsilon_{In^-}\alpha_1)\,c_{In}$$

$$= [(4.0 \times 10^2)(0.511) + (4.51 \times 10^3)(0.489)](1.25 \times 10^{-4}) = \underline{\underline{0.301}}$$

Similarly for part (c)

	pH	$[H_3O^+]$	α_0	α_1	Absorbance
(a)	5.30	5.01×10^{-6}	0.511	0.489	0.301
(c)	6.10	7.94×10^{-7}	0.142	0.858	0.491

23-31. **(a)** The results are shown in the spreadsheet that follows.

(b) The equation is also shown in the spreadsheet.

(c) Here we calculate the standard deviation in the slope, s_m by Equation 7-16

$$s_m = \sqrt{\frac{s_r^2}{S_{xx}}}$$

where s_r is the standard error in y (STEYX). The standard deviation in the intercept, s_b is given by Equation 7-17

$$s_b = s_r\sqrt{\frac{1}{N - (\Sigma x_i)^2 / \Sigma x_i^2}}$$

The results of these calculations are shown in the spreadsheet.

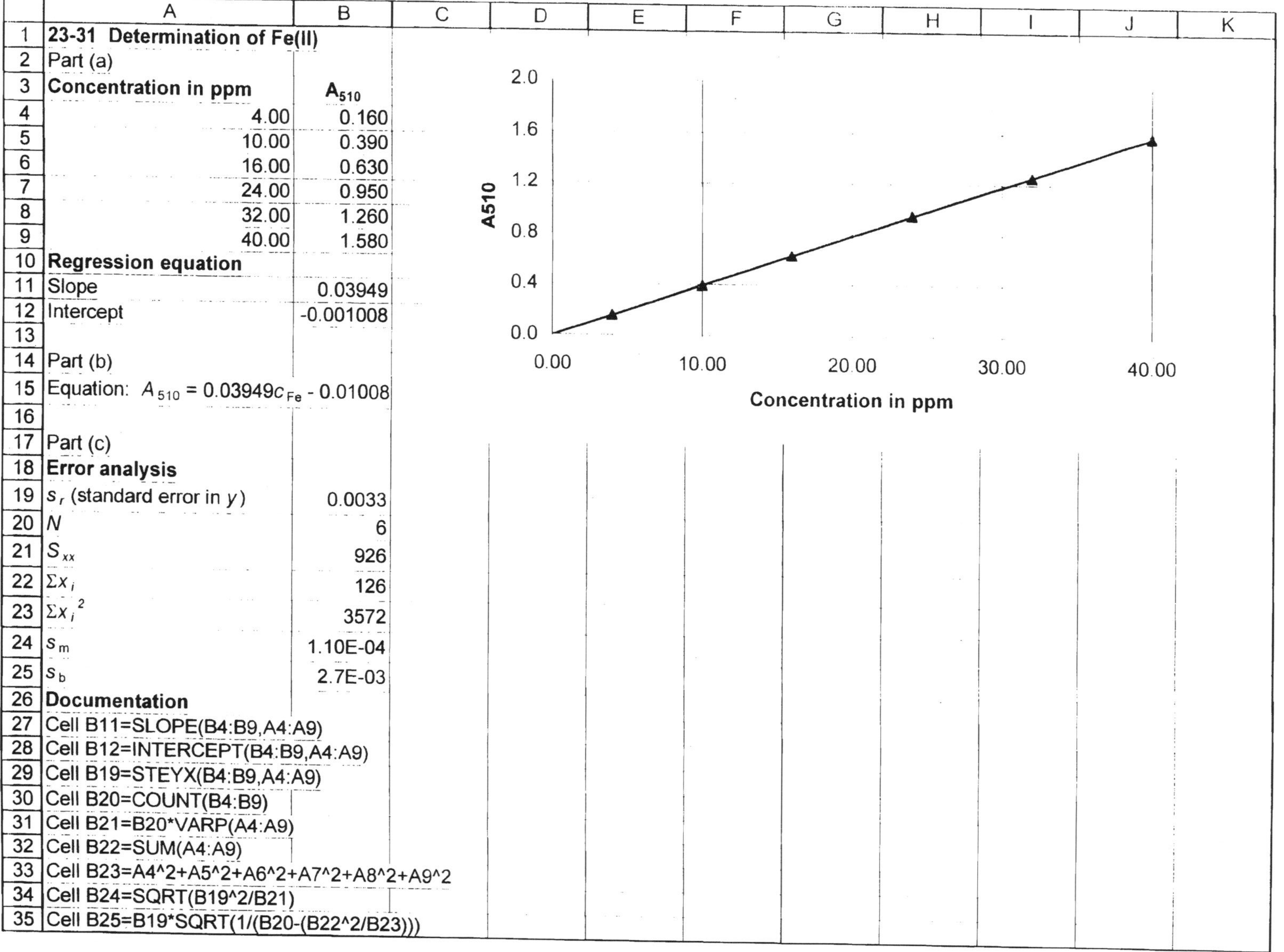

	A	B
1	**23-31 Determination of Fe(II)**	
2	Part (a)	
3	**Concentration in ppm**	**A_{510}**
4	4.00	0.160
5	10.00	0.390
6	16.00	0.630
7	24.00	0.950
8	32.00	1.260
9	40.00	1.580
10	**Regression equation**	
11	Slope	0.03949
12	Intercept	-0.001008
13		
14	Part (b)	
15	Equation: $A_{510} = 0.03949c_{Fe} - 0.01008$	
16		
17	Part (c)	
18	**Error analysis**	
19	s_r (standard error in y)	0.0033
20	N	6
21	S_{xx}	926
22	Σx_i	126
23	Σx_i^2	3572
24	s_m	1.10E-04
25	s_b	2.7E-03
26	**Documentation**	
27	Cell B11=SLOPE(B4:B9,A4:A9)	
28	Cell B12=INTERCEPT(B4:B9,A4:A9)	
29	Cell B19=STEYX(B4:B9,A4:A9)	
30	Cell B20=COUNT(B4:B9)	
31	Cell B21=B20*VARP(A4:A9)	
32	Cell B22=SUM(A4:A9)	
33	Cell B23=A4^2+A5^2+A6^2+A7^2+A8^2+A9^2	
34	Cell B24=SQRT(B19^2/B21)	
35	Cell B25=B19*SQRT(1/(B20-(B22^2/B23)))	

	A	B	C	D	E	F	G	H
1	**23-32 Determination of Fe(II)**							
2								
3	**Concentration in ppm**	**A_{510}**	**Absorbance of unknowns**			**Conc. Fe, ppm**	**s_c 1 result, rel %**	**s_c 3 results, rel %**
4	4.00	0.160	(a)	0.143		3.65	2.8	2.1
5	10.00	0.390	(b)	0.675		17.1	0.54	0.36
6	16.00	0.630	(c)	0.068		1.75	6.1	4.6
7	24.00	0.950	(d)	1.009		25.6	0.36	0.24
8	32.00	1.260	(e)	1.512		38.3	0.27	0.20
9	40.00	1.580	(f)	0.546		13.9	0.68	0.46
10	**Regression equation**							
11	Slope	0.03949						
12	Intercept	-0.001008		**See Problem 23-31 for the calibration curve.**				
13	Equation: $A_{510} = 0.03949c_{Fe} - 0.01008$							
14	**Error analysis**							
15	s_r (standard error in y)	0.0033						
16	N	6						
17	S_{xx}	926						
18	y bar (average absorbance)	0.828						
19	M for single measurements	1						
20	M for replicate measurements	3						
21	**Documentation**							
22	Cell B11=SLOPE(B4:B9,A4:A9)							
23	Cell B12=INTERCEPT(B4:B9,A4:A9)							
24	Cell B15=STEYX(B4:B9,A4:A9)							
25	Cell B16=COUNT(B4:B9)							
26	Cell B17=B16*VARP(A4:A9)							
27	Cell B18=AVERAGE(B4:B9)							
28	Cell B19=1 (entry)							
29	Cell B20=3 (entry)							
30	Cell F4=(D4-$B$12)/$B$11							
31	Cell G4=(($B$15/$B$11*SQRT(1/$B$19+1/$B$16+(D4-$B$18)^2/($B$11^2*$B$17)))/F4)*100							
32	Cell H4=(($B$15/$B$11*SQRT(1/$B$20+1/$B$16+(D4-$B$18)^2/($B$11^2*$B$17)))/F4)*100							

23-33. Q = quinine

$$\text{ppm Q in diluted sample} = 100\text{ ppm} \times \frac{288}{180} = 160.0\text{ ppm} = 160\text{ mg Q/1000 mL}$$

$$\frac{160.0\text{ mg Q}}{10^3\text{ mL soln}} \times 100\text{ ml soln} \times \frac{500\text{ mL}}{15\text{ mL}} = \underline{\underline{533\text{ mg Q}}}$$

	A	B	C	D	E
1	**23-34 Determination of quinine in a tablet**				
2	Initial volume, L	1.000			
3	Dilution factor	5.00			
4	F_{unk}	540			
5	F_{unk+S}	600			
6	Conc. standard, ppm	50			
7	C_S (after dilution)	5			
8	Initial weight sample, g	2.196			
9					
10	Equations	$F_{unk} = kC_{unk}$			
11		$F_{unk+S} = k(C_{unk} + C_S) = kC_{unk} + kC_s$			
12					
13		$F_{unk+S} = F_{unk} + kC_s$			
14		$\therefore k = (F_{unk+S} - F_{unk})/C_s$			
15					
16	and	$C_{unk} = F_{unk}/[(F_{unk+S} - F_{unk})/C_s]$			
17					
18	C_{unk}, ppm	45			
19	C_{unk} in original 1.000 L sample, ppm	225	= 225 mg quinine in the tablet		
20	ppm quinine in tablet	102459	or 10.2%		
21					
22	**Documentation**				
23	Cell B18=$B$4/(($B$5-$B$4)/$B$7)				
24	Cell B19=$B$3*B18				
25	Cell B20=225*1000/$B$8				

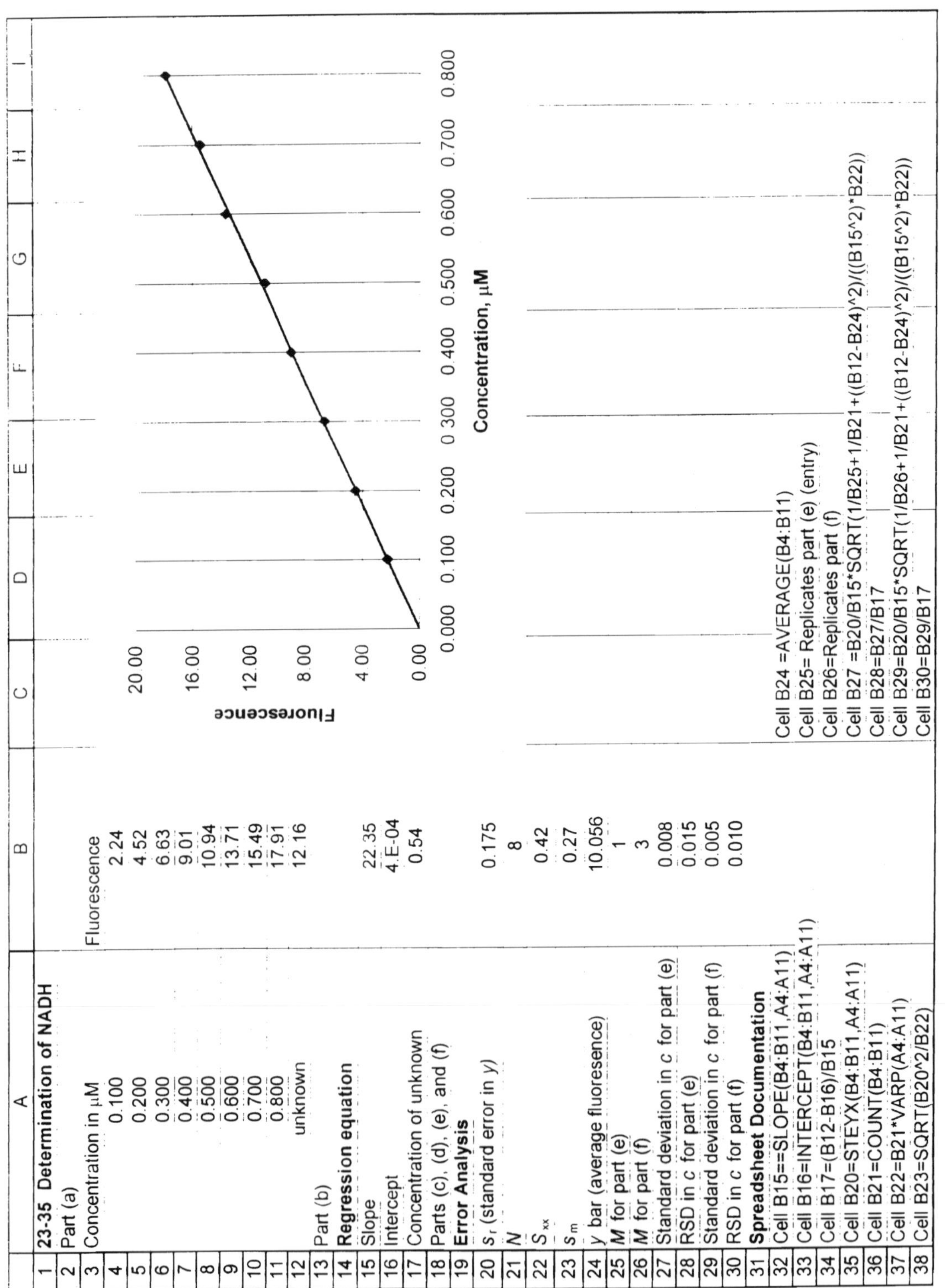

	A	B	C
1	**23-35 Determination of NADH**		
2	Part (a)		
3	Concentration in μM	Fluorescence	
4	0.100	2.24	
5	0.200	4.52	
6	0.300	6.63	
7	0.400	9.01	
8	0.500	10.94	
9	0.600	13.71	
10	0.700	15.49	
11	0.800	17.91	
12	unknown	12.16	
13	Part (b)		
14	**Regression equation**		
15	Slope	22.35	
16	Intercept	4.E-04	
17	Concentration of unknown	0.54	
18	Parts (c), (d), (e), and (f)		
19	**Error Analysis**		
20	s_r (standard error in *y*)	0.175	
21	*N*	8	
22	S_{xx}	0.42	
23	s_m	0.27	
24	*y* bar (average fluoresence)	10.056	
25	*M* for part (e)	1	
26	*M* for part (f)	3	
27	Standard deviation in *c* for part (e)	0.008	
28	RSD in *c* for part (e)	0.015	
29	Standard deviation in *c* for part (f)	0.005	
30	RSD in *c* for part (f)	0.010	
31	**Spreadsheet Documentation**		
32	Cell B15==SLOPE(B4:B11,A4:A11)		Cell B24 =AVERAGE(B4:B11)
33	Cell B16=INTERCEPT(B4:B11,A4:A11)		Cell B25= Replicates part (e) (entry)
34	Cell B17=(B12-B16)/B15		Cell B26=Replicates part (f)
35	Cell B20=STEYX(B4:B11,A4:A11)		Cell B27 =B20/B15*SQRT(1/B25+1/B21+((B12-B24)^2)/((B15^2)*B22))
36	Cell B21=COUNT(B4:B11)		Cell B28=B27/B17
37	Cell B22=B21*VARP(A4:A11)		Cell B29=B20/B15*SQRT(1/B26+1/B21+((B12-B24)^2)/((B15^2)*B22))
38	Cell B23=SQRT(B20^2/B22)		Cell B30=B29/B17

Chapter 24

24-1. A *masking agent* is a complexing reagent that reacts selectively with one or more components of a solution to prevent them from interfering in an analysis.

24-3. $K = c_{org}/c_{aq}$

Applying Equation 24-2 gives

(a)

$$(c_{aq})_n = \left(\frac{V_{aq}}{V_{org}K + V_{aq}}\right)^n (c_{aq})_0$$

$$(c_{aq})_1 = \left(\frac{50.0}{40.0 \times 9.6 + 50.0}\right)^1 0.150 = \underline{\underline{1.73 \times 10^{-2}\ M}}$$

(b)

$$(c_{aq})_2 = \left(\frac{50.0}{20.0 \times 9.6 + 50.0}\right)^2 0.150 = \underline{\underline{6.40 \times 10^{-3}\ M}}$$

(c)

$$(c_{aq})_4 = \left(\frac{50.0}{10.0 \times 9.6 + 50.0}\right)^4 0.150 = \underline{\underline{2.06 \times 10^{-3}\ M}}$$

(d)

$$(c_{aq})_8 = \left(\frac{50.0}{5.0 \times 9.6 + 50.0}\right)^8 0.150 = \underline{\underline{6.89 \times 10^{-4}\ M}}$$

24-5. **(a)** Substituting into Equation 24-2

$$1.00 \times 10^{-4} = \left(\frac{25.0}{25.0 \times 9.6 + 25.0}\right)^n 0.0500 = (0.0943)^n\ 0.0500$$

$$(0.0943)^n = 1.00 \times 10^{-4}/0.0500 = 2.00 \times 10^{-3}$$

Taking the log of both sides of this equation gives

$$n \log 0.0943 = \log 2.00 \times 10^{-3}$$

$$n = \frac{\log 2.00 \times 10^{-3}}{\log 0.0943} = 2.63 = 3$$

$$\text{total volume} = 3 \times 25.0 = \underline{\underline{75\ \text{mL}}}$$

(b) $$1.00\times10^{-4} = \left(\frac{25.0}{10.0\times9.6+25.0}\right)^n 0.0500 = (0.2066)^n\, 0.0500$$

$$n\log 0.2066 = \log 2.00\times10^{-3}$$

$$n = 3.94 = 4 \text{ extractions}$$

$$\text{vol} = 4\times10.0 = \underline{\underline{40\text{ mL}}}$$

(c) $n = 10.9 = 11$ and vol = $\underline{\underline{22\text{ mL}}}$

24-7. (a) $$(c_{aq})_2 = \left(\frac{50.0}{25.0\,K+50.0}\right)^2 (c_{aq})_0$$

$$(c_{aq})_2/(c_{aq})_0 = 1.000-0.990 = 0.0100$$

$$0.0100 = \left(\frac{50.0}{25.0\,K+50.0}\right)^2$$

$$\sqrt{0.0100} = 0.100 = 50.0/(25.0\,K+50.0)$$

$$K = (50.0-50.0\times0.100)/(25.0\times0.100) = \underline{\underline{18.0}}$$

(b) In the same way, $K = \underline{\underline{7.56}}$

24-8. (a) $$(c_{aq})_4 = 0.0500\times1.00\times10^{-6} = 5.00\times10^{-8}$$

$$5.00\times10^{-8} = \left(\frac{30.0}{10.0\,K+30.0}\right)^4\times0.0500$$

$$(1.00\times10^{-6})^{1/4} = 30.0/(10.0\,K+30.0) = 3.16\times10^{-2}$$

$$K = \frac{(30.0-3.16\times10^{-2}\times30.0)}{0.316} = \underline{\underline{91.9}}$$

24-9. **(a)** Assume that in the presence of $HClO_4$, HA is not dissociated to an appreciable extent, whereas in the presence of NaOH, dissociation is complete. From the data for Solution 1, we write

$$K = [HA]_{org}/[HA]_{aq} = 0.0454/\left(0.150\times\frac{50.00}{100.0}-0.0454\right) = \underline{\underline{1.53}}$$

(b) $[HA]_{org} = 0.0225$

$$[HA]_{aq} = 0.0225/1.53 = \underline{\underline{0.0147}}$$

$$[HA]_{org} + [HA]_{aq} + [A^-] = 50.0\times 0.150/100.0 = 0.0750$$

$$0.0225 + 0.0147 + [A^-]_{aq} = 0.0750$$

$$[A^-]_{aq} = \underline{\underline{0.0378}}$$

(c) $[H_3O^+]_{aq} = [A^-]_{aq} = 0.0378$

$$K_a = 0.0378\times 0.0378/0.0147 = \underline{\underline{9.7\times 10^{-2}}}$$

24-11. **(a)**

$$15.3\text{ mL NaOH}\times 0.0202\frac{\text{mmol NaOH}}{\text{mL NaOH}}\times\frac{1\text{ meq cation}}{\text{mmol NaOH}}\times\frac{1}{25.00\text{ mL sample}}\times\frac{1000\text{ mL sample}}{\text{L sample}} = 12.36 = \underline{\underline{12.4\text{ meq cation/L sample}}}$$

(b)

$$12.36\frac{\text{meq cation}}{\text{L sample}}\times\frac{1\text{ mmol CaCO}_3}{2\text{ meq cation}}\times 100.09\frac{\text{mg CaCO}_3}{\text{mmol CaCO}_3} = \underline{\underline{6.19\times 10^2\frac{\text{mg CaCO}_3}{\text{L sample}}}}$$

24-13.

$$2.000\text{ L HCl}\times 0.1500\frac{\text{mol HCl}}{\text{L}}\times\frac{1\text{ mol NaCl}}{\text{mol HCl}}\times\frac{58.442\text{ g NaCl}}{\text{mol NaCl}} = \underline{\underline{17.53\text{ g NaCl}}}$$

Dissolve 17.53 g of NaCl in about 100 mL of water and pass the solution through a column packed with a cation exchange resin in its acid form. Wash the column with several hundred milliliters of water, collecting the liquid from the original solution and the washings in a 2-L volumetric flask. Dilute to the mark and mix well.

24-15. **(a)** *Elution* is a process in which species are washed through a chromatographic column by additions of fresh solvent.

(c) The *stationary phase* in a chromatographic column is a solid or liquid that is fixed in place. A mobile phase then passes over or through the stationary phase.

(e) The *retention time* for an analyte is the time interval between its injection onto a column and the appearance of its peak at the other end of the column.

(g) The *selectivity factor* α of a column toward two species is given by the equation $\alpha = K_B / K_A$ where K_B is the partition ratio of the more strongly held species B and K_A is the corresponding ratio for the less strongly held solute A.

24-17. In gas-liquid chromatography, the mobile phase is a gas, whereas in liquid-liquid chromatography it is a liquid.

24-19. The number of plates in a column can be determined by measuring the retention time t_R and width of a peak at its base W. The number of plates N is then given by the equation $N = 16\,(t_R / W)^2$.

24-21. $N = 16\,(t_R / W)^2$ (Equation 24 – 18)

(a)

	N	N^2
A	$16\,(5.4/0.41)^2 = 2775$	7.7006×10^6
B	$16\,(13.3/1.07)^2 = 2472$	6.1108×10^6
C	$16\,(14.1/1.16)^2 = 2363$	5.5838×10^6
D	$16\,(21.6/1.72)^2 = 2523$	6.3655×10^6
	$\Sigma N = 10134$	$\Sigma N^2 = 25.7607 \times 10^6$

(b) $\overline{N} = 10134/4 = 2534 = \underline{\underline{2.5 \times 10^3}}$

$$s = \sqrt{\frac{25.7607 \times 10^6 - (10134)^2/4}{4-1}} = 170 = 0.2 \times 10^3$$

$$\overline{N} = \underline{\underline{2.5\,(\pm 0.2) \times 10^3}}$$

(c) $H = 24.7\text{ cm}/2534\text{ plates} = 9.747 \times 10^{-3} = \underline{\underline{0.0097\text{ cm}}}$

24-22. (a) $k = (t_R - t_M)/t_M$ (Equation 24 – 12)

A $k_A = (5.4 - 3.1)/3.1 = 0.742 = \underline{\underline{0.74}}$

B $k_B = (13.3 - 3.1)/3.1 = 3.29 = \underline{\underline{3.3}}$

C $k_C = (14.1 - 3.1)/3.1 = 3.55 = \underline{\underline{3.5}}$

D $k_D = (21.6 - 3.1)/3.1 = 5.97 = \underline{\underline{6.0}}$

(b) Rearranging Equation 24-10 yields

$$K_A = k_A V_M / V_S$$

Substituting the values for k from part (a) and the numerical data for V_M and V_S gives

$$K = [(t_R - t_M)/t_M]\ 1.37/0.164 = k \times 8.35$$

and

$$K_A = 0.742 \times 8.35 = \underline{\underline{6.2}}$$

$$K_B = 3.29 \times 8.35 = \underline{\underline{27}}$$

$$K_C = 3.55 \times 8.35 = \underline{\underline{30}}$$

$$K_D = 5.97 \times 8.35 = \underline{\underline{50}}$$

24-23. $R_S = 2\,[(t_R)_C - (t_R)_B]/(W_B + W_C)$ (Equation 24 – 19)

(a) $R_S = 2(14.1 - 13.3)/(1.07 + 1.16) = 0.717 = \underline{\underline{0.72}}$

(b) $\alpha_{C,B} = \dfrac{(t_R)_C - t_M}{(t_R)_B - t_M} = \dfrac{14.1 - 3.1}{13.3 - 3.1} = 1.08 = \underline{\underline{1.1}}$ (Equation 24 – 15)

(c) Proceeding as in Example 24-2(d), page 660, we write

$$\frac{(R_S)_1}{(R_S)_2} = \frac{\sqrt{N_1}}{\sqrt{N_2}} = \frac{0.717}{1.5} = \frac{\sqrt{2534}}{\sqrt{N_2}}$$

$$N_2 = 2534 \times (1.5)^2/(0.717)^2 = 11090 = 1.11 \times 10^4$$

From Solution 24-7(c), $H = 9.75 \times 10^{-3}$ cm/plate

$$L = 11090 \times 9.75 \times 10^{-3} = \underline{\underline{108 \text{ cm}}}$$

(d) Proceeding as in Example 24-2(e),

$$\frac{(t_R)_1}{(t_R)_2} = \frac{(R_S)_1^2}{(R_S)_2^2} = \frac{14.1}{(t_R)_2} = \frac{(0.717)^2}{(1.5)^2}$$

$$(t_R)_2 = 14.1\,(1.5)^2/(0.717)^2 = 61.7 = \underline{\underline{62 \text{ min}}}$$

24-24. (a) From Equation 24-19

$$R_S = 2[(t_R)_D - (t_R)_C]/(W_D + W_C)$$

$$= 2(21.6 - 14.1)/(1.72 + 1.16) = 5.21 = \underline{\underline{5.2}}$$

(b) Proceeding as in part (c) of Solution 24-23 we write

$$N_1 = N_2 \frac{(R_S)_1^2}{(R_S)_2^2} = 2534 \times \frac{(1.5)^2}{(5.21)^2} = 210 \text{ plates}$$

$$L = 210 \text{ plates} \times 9.75 \times 10^{-3} \text{ cm/plate} = \underline{\underline{2.0 \text{ cm}}}$$

24-28. (a) $k_1 = (10.0 - 1.9)/1.9 = 4.26 = \underline{\underline{4.3}}$ (Equation 24 – 12)

$$k_2 = (10.9 - 1.9)/1.9 = 4.74 = \underline{\underline{4.7}}$$

$$k_3 = (13.4 - 1.9)/1.9 = 6.05 = \underline{\underline{6.1}}$$

(b) Rearranging Equation 24-10 and substituting numerical values for V_M and V_S gives

$$K_1 = 4.26 \times 62.6/19.6 = 13.60 = \underline{\underline{14}}$$

$$K_2 = 4.74 \times 62.6/19.6 = 15.1 = \underline{\underline{15}}$$

$$K_3 = 6.05 \times 62.6/19.6 = 19.3 = \underline{\underline{19}}$$

(c) $\alpha_{2,1} = (10.9 - 1.9)/(10.0 - 1.9) = \underline{\underline{1.11}}$ (Equation 24 – 15)

24-29. (a) $k_M = K_S V_S / V_M = 6.01 \times 0.422 = \underline{\underline{2.54}}$ (Equation 24 – 10)

$$k_N = 6.20 \times 0.422 = \underline{\underline{2.62}}$$

(b) $\alpha = 6.20/6.01 = \underline{\underline{1.03}}$

(c) Substituting into Equation 24-21 gives

$$N = 16(1.5)^2\left(\frac{1.03}{1.03-1.00}\right)^2\left(\frac{1.00+2.62}{2.62}\right)^2 = \underline{\underline{8.1\times10^4 \text{ plates}}}$$

(d) $L = 8.1\times10^4\times2.2\times10^{-3} = 178 = \underline{\underline{1.8\times10^2 \text{ cm}}}$

(e) Substituting into Equation 24-22

$$(t_R)_N = \frac{16(1.5)^2\times2.2\times10^{-3}}{7.10}\left(\frac{1.03}{1.03-1.00}\right)^2\frac{(3.62)^3}{(2.62)^2} = \underline{\underline{91 \text{ min}}}$$

Chapter 25

25-1. *Slow sample injection* in gas chromatography leads to band broadening and lowered resolution.

25-3. A *chromatogram* is a plot of detector response, which is proportional to analyte concentration, as a function of time. Since the detector is positioned at the end of a column the chromatogram shows peaks, or bands, of various analytes as they emerge from a chromatographic column.

25-5. In *open tubular columns*, the stationary phase is held on the inner surface of a capillary, whereas in packed columns, the stationary phase is supported on particles that are contained in a glass or metal tube. Open tubular columns, which are only applicable in gas and supercritical fluid chromatography, contain an enormous number of plates that permit rapid separations of closely related species. They suffer from small sample capacities.

25-7. **(a)** diethyl ether, benzene, *n*-hexane

25-8. **(a)** ethyl acetate, dimethylamine, acetic acid

25-9. In *adsorption chromatography*, separations are based upon *adsorption equilibria* between the components of the sample and a solid surface. In partition chromatography, separations are based upon *distribution equilibria* between two immiscible liquids.

25-11. *Gel filtration* is a type of size-exclusion chromatography in which the packings are hydrophilic, and eluents are aqueous. It is used for separating high-molecular-weight polar compounds. *Gel-permeation chromatography* is a type of size-exclusion chromatography in which the packings are hydrophobic, and eluents are nonaqueous. It is used for separating high-molecular-weight nonpolar species.

25-13 Quantitative GC of 3 Components by Area Normalization

Component No.	Area	Rel. Response	Corrected	Percentage
1	16.4	0.60	27.33	22.9
2	45.2	0.78	57.95	48.5
3	30.2	0.88	34.32	28.7
		Total Area	119.60	

Spreadsheet Documentation

Cell D4=B4/C4

Cell D5=B5/C5

Cell D6=B6/C6

Cell D7=SUM(D4:D6)

Cell E4=(D4/$D$7)*100

Cell E5=(D5/$D$7)*100

Cell E6=(D6/$D$7)*100

25-16. The simplest type of pump for liquid chromatography is a *pneumatic pump*, which consists of a collapsible solvent container housed in a vessel that can be pressurized by a compressed gas. This type of pump is simple, inexpensive, and pulse-free. It has limited capacity and pressure output, it is not adaptable to gradient elution, and its pumping rate depends upon the viscosity of the solvent.

A *screw-driven syringe pump* consists of a large syringe in which the piston is moved in or out by means of a motor-driven screw. It also is pulse-free and the rate of delivery is easily varied. It suffers from lack of capacity and is inconvenient to use when solvents must be changed.

The most versatile and widely used pump is the *reciprocating pump* that usually consists of a small cylindrical chamber that is filled and then emptied by the back-and-forth motion of a piston. Advantages of the reciprocating pump include small internal volume, high output pressures, adaptability to gradient elution, and flow rates that are constant and independent of viscosity and back pressure. The main disadvantage is pulsed output that must be damped.

Chapter 26

26-1. The properties of a supercritical fluid that are of particular importance to its application to chromatography are its density, its viscosity, and the rates at which solutes diffuse in it. The magnitude of each of these properties lies intermediate between a typical gas and a typical liquid.

26-3. Pressure increases cause the density of a super critical fluid to increase, which causes the k for analytes to change. Generally increases in pressure reduce the retention times of solutes.

26-5. Supercritical fluid chromatography is particularly applicable to nonvolatile and thermally unstable compounds that contain no chromophoric functional groups.

26-7. **(a)** An increase in flow rate results in a decrease in retention time.

(b) An increase in pressure results in a decrease in retention time.

(c) An increase in temperature results in a decrease in density of supercritical fluids and thus an increase in retention time.

26-9. Electroosmotic flow can be repressed by reducing the charge on the interior of the capillary by chemical treatment of the surface.

26-11. Under the influence of an electrical field mobile ions in a solution are attracted or repelled by the negative potential of one of the electrodes. The rate of movement toward or away from a negative electrode is dependent upon the net charge on the analyte and the size and shape of the analyte molecules. These properties vary from analyte to analyte. Thus the rate at which molecules migrate under the influence of the electrical field vary, and the time it takes them to traverse the column varies, thus making separations possible.

26-13. The migration rate v, is given by Equation 26-1:

$$v = \mu_e \cdot \frac{V}{L} = 4.31 \times 10^{-4}(\text{cm}^2\,\text{s}^{-1}\,\text{V}^{-1}) \times \frac{10 \times 10^3\ \text{V}}{50\ \text{cm}}$$

$$= 0.0862\ \text{cm s}^{-1}$$

The electroosmotic flow rate $= 0.085\ \text{cm s}^{-1}$

The rate of movement of a cation $= 0.0862 + 0.085 \quad = \quad 0.1712 \text{ cm s}^{-1}$

$$\text{Time required} = \frac{40 \text{ cm}}{0.1712 \text{ cm s}^{-1}} = 234 \text{ s} \quad \text{or} \quad \frac{234}{60} = 3.9 \text{ min}$$

26-15. The advantages of micellar electrokinetic capillary chromatography over HPLC are higher column efficiencies and the ease with which the pseuduostationary phase can be altered.